CECO
195 ELIZ
GANDER, N
TEL: 709-256-7112
FAX: 709-256-8324

Challenge and Change

An Illustrated History of Engineering and Geoscience in Newfoundland and Labrador

by
D. R. Tarrant, P. Eng

Association of Professional Engineers and Geoscientists of Newfoundland
St. John's, Newfoundland and Labrador. Canada
2002

Challenge and Change

An Illustrated History of Engineering and Geoscience in Newfoundland and Labrador

Commemorating fifty years of Professional Engineering and Geoscience in Newfoundland and Labrador (1952 - 2002)

National Library of Canada Cataloguing in Publication Data

Tarrant, D. R. (Donald R.), 1946-
    Challenge and change: an illustrated history of engineering and geoscience in Newfoundland and Labrador / D.R. Tarrant.

Includes bibliographical references and index.
ISBN 0-9731777-0-5

    1. Engineering--Newfoundland and Labrador--History.
    2. Geology--Newfoundland and Labrador--History. I. Association of Professional Engineers and Geoscientists of Newfoundland. II. Title.

TA27.N6T37 2002   620'.009718   C2002-904777-3

Copyright © 2002 by D. R. Tarrant

ALL RIGHTS RESERVED. No part of the work covered by the copyright hereon may be reproduced or used in any form or by any means—graphic, electronic or mechanical—without the written permission of the author. Any request for photocopying, recording, taping or information storage and retrieval systems of any part of this book shall be directed to the Canadian Reprography Collective, 379 Adelaide Street West, Suite M1, Toronto, Ontario M5V 1S5. This applies to classroom use as well.

PRINTED IN CANADA BY ROBINSON-BLACKMORE

Association of Professional Engineers
and Geoscientists of Newfoundland
P. O. Box 21207, St. John's, NF, Canada, A1A 5B2
Telephone: (709) 753-7714, Facsimile (709) 753-6131
E-mail main@apegn.nf.net

Table of Contents

Acknowledgements
Foreword - President of APEGN

1	Early Engineering in Newfoundland	1
2	Communications - From Gisborne to Satellites	9
3	Electricity - From Petty Harbour to the Mighty Churchill	51
4	Transportation - Land, Sea and Air	83
5	Mining - From Ballast to Bullion	119
6	Pulp and Paper - Serving the Newspapers of the World	155
7	Construction	169
8	Marine - Fishery and Shipbuilding	189
9	Oil and Gas - From Parsons Pond to Hibernia and Beyond	201
10	Engineering and Geoscience Education and Research in the Province	221
11	The History of APEGN	231
12	Epilogue	241
	Appendix	243
	Bibliography of Works Consulted	259
	Index	263

APEGN gratefully acknowledges the generous financial support for the publishing of this book from the following firms:

Abitibi Consolidated - Grand Falls Division
Acres International Limited
Aliant Telecommunications Inc.
Amec E&C Services Ltd.
Campbell Engineering Ltd
Canadian Offshore Investments Ltd.
City Paving Ltd.
Coflexip Stena Offshore Newfoundland Limited
Consulting Engineers of Newfoundland & Labrador
Controls & Equipment Ltd.
Crosbie Engineering Ltd.
Custom Systems Electronics Limited
DRM Consulting
Eastern Technical Services
FGA Consulting Engineers
First Copy Duplicating Centre Ltd.
Group Telecom
H. T. Kendall & Associates Ltd.
Harris & Associates Limited
Info Tech Engineering
Jacques Whitford Group & Nfld Geosciences
Kavanagh & Associates Limited
Marlyn Construction Ltd.
McNamara Construction Company
Modern Paving Ltd.
Nadoo Engineering
Newfoundland Power Inc.
Newlab Engineering Ltd.
Newton Engineering Ltd.
Noble Offshore Ltd.
North Atlantic Refining Limited
Nova Consultants Inc.
NSB Offshore Inc.
Petro-Canada
Project Management Services
Provincial Consultants Ltd.
Provincial Refrigeration Ltd.
Quadratec Inc.
R. Mercer Exploration & Development
Redwood Construction
Schlumberger
SGE Group Inc.
Smith Stockley Ltd.
Structural Design Inc.
Trico Limited
Trident Construction Ltd.
Voisey's Bay Nickel Company Ltd.

Acknowledgements

I would like to acknowledge with sincere appreciation the help, advice and assistance I received in preparing *Challenge and Change*. Thanks go to the staffs of the Provincial Archives of Newfoundland and Labrador, the Newfoundland Collection of the A. C. Hunter Library, the Memorial University of Newfoundland Centre for Newfoundland Studies and Archives, the Memorial University of Newfoundland Maritime History Archive, and the City of St. John's Archives. Thanks also go to the many corporations, associations and individuals who provided information and photographs for this volume. Special thanks go to those who advised me on the manuscript including Dr. Ross Peters, P. Eng., Professor Howard Dyer, P. Eng., Dr. Wallace Read, P. Eng., John Evans, P. Eng., Dr. Hugh Miller, P. Geo., Willis Martin, P. Eng., and my wife Dr. Sandra Clarke.

Special thanks go to Steve McLean, P. Eng who organized this project and helped bring the book to a successful conclusion. A particular thank-you is extended to the engineering corporations who financially supported this publication.

Writing a book on Newfoundland and Labrador's engineering and geoscience legacy was indeed a daunting task. I trust that *Challenge and Change* demonstrates the many challenges that engineers and geoscientists have overcome in their contributions to development and change in the province.

Donald R. Tarrant P. Eng.
St. John's, Newfoundland and Labrador, Canada
October, 2002

Photo Credits

A. C. Hunter Library, Newfoundland Collection
Abitibi-Consolidated
Association of Professional Engineers and Geoscientists of Newfoundland (APEGN)
Baine Johnston Corporation (Baine Johnston)
Canadian Coast Guard
City of St. John's Archives
Corner Brook Pulp and Paper (CBPP)
Department of Works, Services and Transportation
The Express
Fishery Products International (FPI)
George Neary
Grand Falls-Windsor Heritage Society
Hibernia Management Development Corporation (HMDC)
Iron Ore Company of Canada (IOC)
Larry Sheehy
McNamara Construction
Mount Pearl Admiralty House Museum
Memorial University Photographic Services (MUN Photographic Services)
Memorial University Centre for Newfoundland Studies
    Archives (MUN CNS Archives)
Memorial University Map Library (MUN Map Library)
Memorial University Maritime History Archive (MUN Maritime History Archive)
National Research Council - Institute for Marine Dynamics (NRC-IMD)
Newfoundland and Labrador Hydro-electric Corporation (NLH)
Newfoundland Historical Society (NHS)
Newfoundland Transshipment Limited
Newfoundland Power (NP)
Newtel
North Atlantic Refining
Peter Kiewit Sons
Petro-Canada
Public Archives of Newfoundland and Labrador (PANL)
Red Indian Lake Development Association
Roger Angel, P. Eng.
Roger Jamieson
Port Authority of St. John's
The Telegram
VOWR
Wabush Mines

Foreword

On June 7, 2002 the Association of Professional Engineers and Geoscientists of Newfoundland marked its 50th anniversary. To commemorate that significant milestone the Association commissioned this book on the history of engineering and geoscience in Newfoundland and Labrador.

Since 1952, the practice of engineering – along with the practice of geoscience since 1989 – have been self-regulated by the professional engineers and geoscientists of the Association of Professional Engineers and Geoscientists of Newfoundland, on behalf of the people of the province.

The purposes of regulation of professions and occupations are public protection and the provision of public accountability. For the past fifty years, APEGN has met these purposes for the engineering and geoscience professions in an exemplary fashion, through the dedicated work of the men and women who volunteer their time and energies to our organization.

Engineers and geoscientists create wealth and intellectual property in Newfoundland and Labrador, which are essential to our continued growth. In addition, engineers and geoscientists contribute to the social and political well being of our Province.

Strong engineering and geoscience professions are essential to our continued prosperity. It is, necessary, therefore, that the professions continue to be guided and regulated by a strong and vibrant professional association. We look forward to the future with the knowledge that our professions have much to offer and contribute to our economy and society.

We congratulate Don Tarrant P. Eng. on the completion of this book and thank all who have in any way contributed to its creation and publication.

Charles E. Sheppard, P. Eng.
APEGN President 2001 - 2002

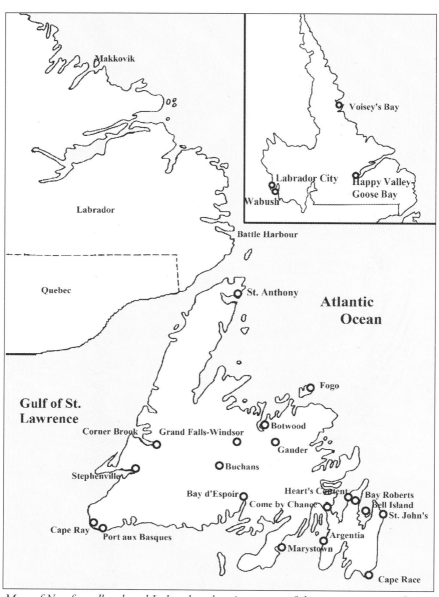

Map of Newfoundland and Labrador showing some of the communities referenced in the text

Chapter One

Early Engineering in Newfoundland

The history of engineering goes back to beyond the great pyramids of Egypt and the hanging gardens of Babylon. In Newfoundland and Labrador, its beginnings were somewhat more humble. When the island's first European settlers arrived, they built houses, boats, wharves, and later, roads – projects that could all be termed works of engineering. In modern use, however, the term has a special meaning denoting the regulated discipline of professional engineering, a vocation which has existed in Newfoundland for only the last half century, with the year 2002 marking the fiftieth anniversary of the Association of Professional Engineers and Geoscientists of Newfoundland. Yet before the arrival of professional engineering and geoscience, many large engineering projects were undertaken in Newfoundland. Some of these have been quite significant, as the following chapters will show.

European settlement did not begin in Newfoundland until the seventeenth century, when John Guy established a colony at Cupids, Conception Bay. As this site is in the early stages of excavation, it will be some years before we understand the full magnitude of that colonization effort. However, we know that the colonists constructed buildings and roadways. In the 1620s, Lord Calvert founded the Colony of Avalon in Ferryland. The remains of this colony have been under excavation for a number of years and the complexity of this venture is now being revealed. The layout of the buildings, streets, and water and sewer systems were works of engineering of the type we now take for granted.

In the early eighteenth century, the British and French built many forts in Newfoundland; the massive forts at Castle Hill in Placentia and

Model of Fort William, St. John's
This British fort had its beginnings in the seventeenth century but after the completion of Fort Townshend in the 1770s, it lost its military importance and was relegated to a barracks for the British garrison until 1871. Part of the structure served as St. John's first railway station. (Courtesy of the City of St. John's Archives)

Fort Townshend in central St. John's are excellent examples of military engineering and construction. The army engineers who designed these structures were probably the first to work in Newfoundland as trained engineering specialists. These forts were complex structures with fortifications, gun batteries, living quarters, and tunnels. In St. John's, forts constructed by the British include Fort William, Fort George, Fort Amherst, and the aforementioned Fort Townshend, some of which were joined by road.

The nineteenth century saw the construction of large churches, especially in the St. John's area. The Church of England St. John the Baptist Cathedral and the Roman Catholic St. John the Baptist Basilica are both massive stone structures providing excellent examples of the engineering and architecture of the period. Other notable buildings of this era were the St. John's Athenaeum, St. Thomas' Church, Government House, and the Colonial Building, not to mention many other churches, residences, merchant buildings, and lighthouses in all parts of the province.

Engineering fully became recognized as a specialty with the advent of such technologies as steam power, electricity, and communications during and after the Industrial Revolution of the late eighteenth and early nineteenth centuries. The technology associated with the production of railroads, electric generating plants, and telegraph facilities was well beyond the capacity of untrained people. Engineers designed and built these complex systems well before the existence of regulatory bodies.

The Canadian scene echoes this situation. Huge engineering efforts such as the Rideau Canal and the transcontinental railway were indeed engineering marvels, but engineering as a profession did not officially begin until 1887, when a group of engineers formed the Canadian Society of Civil Engineers. The year 2002 marks not only the one hundred and fifteenth anniversary of professional engineering in Canada, but also the fiftieth year of professional engineering in Newfoundland and Labrador.

It 1948, the Engineering Institute of Canada (EIC), an association of Canadian professional engineers, opened a branch in St. John's and invited engineers to join. Newfoundland engineers, before Confederation, had no official professional status. EIC members therefore began the process of setting up an association to regulate the profession, and organized the Association of Professional Engineers of Newfoundland (APEN) to administer this function. In May 1952, the Newfoundland Legislature approved an Act giving the association responsibility for overseeing the regulation of engineering in the province. Registered members of APEN would from then on carry the designation Professional Engineer (P. Eng.), a term indicating their competence in a field of engineering, thereby ensuring satisfactory standards and a high level of public safety. We should note that it was "professional engineering" which was officially recognized back in 1952. "Engineering," of course, was being performed in Newfoundland at the time, both by graduate engineers mainly from

Canadian and American universities, as well as by people without formal training. The establishment of APEN ensured that there would be a consistent standard for all engineers in the province. In 1989, the association expanded to include geoscientists, and in recognition of this, changed its name to the Association of Professional Engineers and Geoscientists of Newfoundland (APEGN).

Engineering and geoscience have made tremendous contributions to the province over the past fifty years and have been an integral part of all major construction projects. Even before the formation of APEN, engineers and geoscientists were making an impact. New electrical and mechanical technologies surfaced in the mid 1800s, and the role of engineers and geoscientists changed to harness these new opportunities. Some of the significant engineering projects of the nineteenth and early twentieth centuries were the first trans-island telegraph cable (1856), the first transatlantic telegraph cable (1858), the iron ore mines at Bell Island (1895), the lead and zinc mine at Buchans (1926), the Newfoundland Railway (1898), the first hydro station at Petty Harbour (1900), the paper mills at Grand Falls (1909) and Corner Brook (1925), and the huge wartime military bases (1940s). Also not to be overlooked were the contributions made by engineers to municipal, highway, building, wharf and ship construction.

Fifty years ago in the fields of communications and information technology, dial telephone service was only available in St. John's; radio-telephone was the standard way of communicating with the mainland and Europe; radio broadcasting was only beginning to establish itself; television had not even been introduced to the province; and the transistor, which had been invented only a few years before, was just beginning to find applications. Today state-of-the-art optical fibre systems and satellite links connect Newfoundland and Labrador to North American and world communications networks; just about every desk has a computer terminal with most connected to the internet; and cellphones are almost mandatory for many business uses. Modern information and communications technology has radically transformed the way engineers and geoscientists work, and has undoubtedly had the greatest impact on the professions over the past half century.

With regard to the province's electrical facilities at the time of APEN's formation, several hundred communities still did not have electric power, the province's entire power generating capacity was less than two hundred megawatts, and there was no common power grid (in fact two different frequencies were employed). Since then, the province has constructed the massive Churchill Falls hydro development (at the time the largest civil works project engineered and built in Canada), as well as other large installations including the Bay d'Espoir hydro and the Holyrood thermal generating stations. In the mid 1960s a common Newfoundland power grid was established, the largest non-connected grid in North America. Elsewhere on the continent, a neighbouring province or state could be called upon to make up a power shortfall, an advantage which is not available to the island of Newfoundland, because it is not electrically connected with the mainland.

Transportation and other infrastructures have also undergone massive transformation. Fifty years ago passenger jet service had not even begun, and most of Newfoundland's airports were still being used as military bases. The province's main airports, especially St. John's, have seen tremendous improvements over the past few decades with the introduction of new automated systems to handle today's sophisticated jet aircraft. While the railway system has disappeared, its demise has led to improvements to the province's highway structure. The Trans-Canada Highway, completed between St. John's and Port aux Basques in 1965, has been undergoing constant improvements since then, and is itself a major engineering accomplishment considering the harsh Newfoundland terrain.

In 1952, Newfoundland's largest mine was at Bell Island, and the technology employed was related primarily to ore extraction. This mine, along with others at Buchans and St. Lawrence, has long since disappeared and more modern mines have taken their place. To see the changes in mining technology, one only has to look at the iron mines and processing plants in Labrador City and Wabush. The engineering challenges connected with these western Labrador operations were immense, considering the logistics associated with the remoteness of the area.

Probably the biggest engineering challenge in Newfoundland over the past half century has been the development of our oil and gas resources. Starting with the construction of the oil refineries at Holyrood and Come by Chance, up to the engineering marvels of the Hibernia and Terra Nova projects, the engineering employed has been high-tech and massive in scale, requiring multi-disciplinary approaches. The Hibernia project, for instance, has often been cited as one of the largest engineering projects undertaken in Canada. From an engineering perspective, the construction site of the offshore platforms has an interesting historical significance because it was also at Bull Arm that modern technology first appeared in Newfoundland with the landing of the first transatlantic telegraph cable in 1858.

Over the past fifty years, thousands of engineers and geoscientists have contributed to the province's development. In the early days, most of the engineering for major projects was done outside the province. This was understandable considering the small population and limited number of specialists located here. However, as local engineers and geoscientists gained experience, they came to play a more significant role. In the offshore oil and gas sector, for instance, participation has increased from the early days of the industry to the current White Rose project, for which all of the topsides engineering and project management will be done in Newfoundland and Labrador.

Engineering education in the province has also made tremendous strides. In 1952, Memorial University had two faculty members on its engineering staff and provided a three year pre-engineering program. The Engineering and Applied Science faculty now has approximately fifty professors and more than fourteen hundred students, offering degree programs at all levels and in a range of disciplines. The intervening period has seen the construction of many new facilities and the creation of several research centres, which attract engineering and geoscience scholars from all over the world.

In this book, Newfoundland's engineering and geoscience history will be reviewed by looking at its finished products, such as the power stations, highways, buildings, mines, and oil platforms. Of course, sharing credit for these projects are the other professions which helped construct them; however the engineer's and geoscientist's pride lies in the fact that the projects would not exist without their participation.

Although this volume will focus on the major engineering and geoscience projects in the province, readers must not overlook the not so spectacular routine projects, such as the buildings, highways, and municipal infrastructures of the province's towns, where significant engineering and geoscience contributions have also been made. Professional engineering and geoscience does not always result in megaprojects, but it does contribute to improvements in our everyday life. The professions ensure that citizens can walk under a bridge without fear of its falling down, take an elevator to top of the highest building without fear of either collapsing, take a public service vehicle without concern of its breaking down, or turn on the computer and access internet sites on the other side of the world. Engineering and geoscience is part of everyday life; however, the professions that ensure public safety are seldom the focus of public recognition. The intent of this volume is to shed light on the role of engineering and geoscience in the province, particularly their contribution to provincial development and public safety.

The technical side of engineering and geoscience would obviously be of interest to those working in these particular fields. However, the multitude of projects in the province makes it impossible to focus on the detailed processes, drawings, calculations, and computer simulations that professionals in these fields employ. A single project could easily require an entire volume. The projects themselves are testaments to the engineering and geoscience skills that they required.

In the era following the Industrial Revolution, it was with the efforts to build a telegraph system across the island that the story of engineering and geoscience in Newfoundland really begins. It is this story that is outlined in the next chapter.

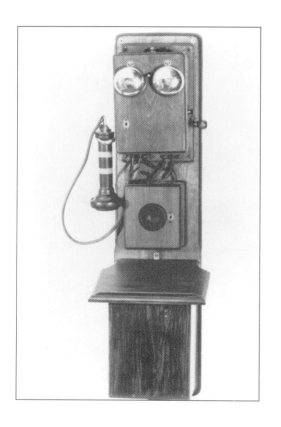 

The magnetophone and the cellphone represent both ends of almost one hundred years of telephone communications. (Courtesy of Newtel)

Chapter Two

Communications - From Gisborne to Satellites

One of the engineering disciplines which had its beginning in the mid nineteenth century was communications engineering. It began with Samuel Morse's invention of the telegraph in 1836 and blossomed as a profession when engineers on both sides of the Atlantic contemplated the notion of a transatlantic telegraph cable. Newfoundland's role in transatlantic communications came about because of its geographic location as the closest part of North America to Europe, making it an ideal western terminus for any transatlantic communications link.

Communications engineering in the province began with the installation of short telegraph links between nearby communities, followed by a much longer system across the island connecting with Nova Scotia. This was later followed by the huge engineering undertaking of the transatlantic telegraph cable that came ashore at Bull Arm in 1858, providing a transatlantic telegraph service for a short time. This cable failed prematurely and transatlantic communication was not re-established until the *Great Eastern* landed a cable at Heart's Content in 1866. Telephones appeared in the latter part of the nineteenth century and the first transatlantic telephone cable was placed almost eighty years later.

Frederick Newton Gisborne (1824-1892)

Gisborne was involved in several aspects of engineering in Newfoundland. Apart from the early military engineers, he was one of the first to practise in Newfoundland. He is best known for introducing telegraphy to the colony in 1851. He was also associated with mining, and explored for minerals in the late 1850s. He represented Newfoundland at the London Exhibition in 1862 and the Paris Exposition of 1863. He left Newfoundland and pursued a number of mining activities in Nova Scotia and British Columbia. Gisborne Lake on the province's south coast is named after him. (Courtesy of PANL)

These transatlantic telegraph and telephone systems were major feats of civil, electrical and mechanical engineering, and also represented some of the earliest projects which employed ocean engineering.

Wireless communications also put Newfoundland in the headlines with Marconi's 1901 transatlantic signal. A year or so afterwards, wireless telegraph stations, primarily for ship to shore communications, began to be installed throughout the island. Wireless telegraphs were followed in the 1920s by wireless voice communications. Radio broadcasting was also introduced in the 1920s, followed by television in the 1950s.

The early communications systems dramatically changed the way people lived. This was particularly the case when telegraphic communications between North America and Europe were established, greatly affecting politics, commerce and social interaction between the two continents.

TELEGRAPH COMMUNICATIONS

Telegraph communication was introduced to Newfoundland by Frederick Newton Gisborne (1824-1892), an English engineer who had previously been active in expanding telegraphy in Quebec and Nova Scotia. Gisborne suggested that if the North American telegraph system were extended to St. John's, transatlantic steamers could stop at the port and drop off messages which then could be relayed by telegraph to North American newspapers more than forty-eight hours earlier than messages arriving on steamers at Halifax or New York. Newspapers at the time paid handsomely for news from Europe, and any competitive advantage would result in a substantial commercial gain.

In September 1851, while still head of the Nova Scotia Telegraph Company, Gisborne petitioned the legislature in St. John's with a plan to construct a telegraph line between St. John's and Cape Ray, which would eventually connect by submarine cable to the Nova Scotia telegraph system. The government approved his application and granted him £500 to survey a route across the island. Prior to tackling this difficult project, he set up the St. John's and Carbonear Electric Telegraph Company to provide telegraph communications between the two towns. Construction of the system was uneventful and service began on March 6, 1852. In the Industrial Revolution era, this was one of the earliest engineering undertakings in Newfoundland, and Gisborne thereby became one of the first engineers to practise in the colony.

Cyrus Field (1819-1892)
Field was indefatigable in his quest to install a transatlantic telegraph cable. After numerous failed attempts, he succeeded in placing a successful cable between Valentia, Ireland and Heart's Content on July 26, 1866. Field crossed the Atlantic more than forty times in this pursuit. (Courtesy of PANL)

TRANS-ISLAND TELEGRAPH LINE

In the spring of 1852, the legislature passed an act incorporating the Newfoundland Electric Telegraph Company (Newfoundland Electric), giving it the right to build telegraph lines between St. John's and Cape Ray, as well as branch lines to Trepassey and other locations. Gisborne organized the company in New York, financed primarily by American investors including Horace Tebbets and Darius B. Holbrook.

Gisborne himself held twenty-five shares. The initial phase of the construction began in the summer of 1853, and employed 350 men. It started at Brigus, where the cable connected into the St. John's to Carbonear system built the year before. After only forty miles of construction, the project was halted when its New York backers, for unknown reasons, failed to honour Gisborne's bills. The Newfoundland Electric Telegraph Company became insolvent and the project came to a halt.

In 1854, Gisborne travelled to New York City to look for new financial support. There he met Cyrus W. Field, a semi-retired merchant who had made a fortune in the paper business. At Field's home, Gisborne expressed his view of the commercial potential of extending the Nova Scotia telegraph system to Newfoundland, and raised the issue of a transatlantic cable. After Gisborne had left, Field pondered Gisborne's ideas and became absorbed with the notion of connecting America and Europe telegraphically. Within a matter of days, he organized a company named the New York, Newfoundland and London Telegraph Company (New York Telegraph), and began work on carrying out this project.

In March 1854, Field travelled to Newfoundland, where he met with Governor Ker Baillie Hamilton and arranged for New York Telegraph to take over Newfoundland Electric's charter. New York Telegraph agreed to complete the telegraph line between St. John's and Cape Ray, and in return was granted a number of concessions, among them a fifty-year monopoly on telegraphic communications in the colony.

In August of 1855, Field attempted to place a submarine cable across the Cabot Strait. The cable arrived from England aboard the *Sarah L. Bryant* and would have been difficult to transfer to a steamer for installation. Samuel Canning, the project's supervising engineer, arrived on the *Bryant* and decided that the *Bryant* would pay out its cable as it was being towed by the steamship *James Adger*.

The installation began smoothly; however, when the ship was only forty miles from Cape Ray, a storm came up forcing the captain to cut the cable to save the ship from sinking. Despite this setback, Field tried another attempt the following year by loading the cable on board the steamship *Propentis*, which successfully laid the cable without incident.

The St. John's to Cape Ray line followed the south coast of the island and logistical support was provided by the *Victoria*, which delivered construction material and supplies to the tops of the bays and inlets along the route. The project employed six hundred men and was an enormous undertaking at the time, considering that prior to this, most projects in Newfoundland had been confined to small local jobs such as buildings or wharves, whereas the cross-island telegraph line extended over a long distance. On October 1, 1856, the telegraph line carried its first message to Baddeck, Cape Breton Island from J. & W. Pitts, a mercantile firm in St. John's.

"Via Cape Race"
In 1859, the Associated Press of New York stationed a boat at Cape Race to intercept transatlantic ocean steamers on their way between North America and Europe. News and messages from Europe were thrown overboard from the steamers in water-tight canisters and were retrieved and telegraphed to North American newspapers from the telegraph office at Cape Race. This continued until the first successful transatlantic cable was installed in 1866 and was immortalized in North American newspapers with the byline "Via Cape Race." (Courtesy of PANL)

TRANSATLANTIC TELEGRAPH CABLES

In London, Field incorporated the Atlantic Telegraph Company to install a telegraph link across the Atlantic. After failed attempts in 1857 and early 1858, the company finally completed a cable between Valentia,

Ireland and Bull Arm (now Sunnyside), Trinity Bay on August 4, 1858. Field's technical advisor on the project was Samuel Morse, the inventor of the telegraph as well as of the internationally known code which bears his name. Morse tested the cables in England by sending signals through a two thousand-mile length of wire submerged in tanks filled with salt water and demonstrated that telegraphy would work over that distance.

For the first time, Europe and North America could communicate with each other telegraphically, and messages between the two continents took minutes rather than days. Despite the euphoria of a transatlantic link, the success of the cable was short-lived as it developed electrical problems and ceased operation after about a three week period, during which only four hundred messages were handled.

The Great Eastern in Heart's Content
*Built in 1858 as a passenger ship, the **Great Eastern** was the largest vessel in the world, but proved to be uneconomical to operate. It was eventually converted into a cable ship. The Great Eastern laid the first two successful transatlantic cables to Heart's Content in July and September, 1866. The ship laid two other cables to Heart's Content in 1873 and 1874. (Courtesy of PANL)*

Heart's Content (Anglo-American Telegraph Company)

Because the United States civil war was in progress, a transatlantic telegraph cable was of secondary concern to British and American investors. It was not until 1865, after the war had finished, that Field again headed to England and set up a new company named Anglo-American Telegraph Company (AAT). To lay the Atlantic cable, AAT procured the *Great Eastern* ocean liner, the largest ship in the world, which was then out of service because it was too costly to operate. The *Great Eastern* was converted to a cable ship, and in 1865 began laying a transatlantic cable from Valentia, Ireland. However, when only several hundred miles short of Heart's Content, the cable snapped. Despite several attempts to recover it, none were successful. The following year, the company made another try to bridge the Atlantic and on July 27, the *Great Eastern* successfully landed a cable at Heart's Content. Unlike the 1858 project, this was the first commercially successful transatlantic link between Europe and North America. Several weeks later, the *Great Eastern* and several escort ships headed back to sea, and after several attempts, recovered the lost 1865 cable. A splice was made and the *Great Eastern* proceeded to Heart's Content, paying out new cable along the way. The second cable was landed at Heart's Content on September 8, thereby providing two transatlantic cable connections within a matter of several weeks. AAT installed other transatlantic cables at Heart's Content in 1873 *(Great Eastern)*, 1874 (*Great Eastern*), 1880 (*Seine*), and 1894 (*Scotia*).

The transatlantic telegraph project was one of the largest engineering undertakings up to that period. The cable's design, its manufacture, and the paying out equipment for its 2000 mile length beneath two miles of ocean depth were engineering issues that had not been previously encountered.

Harbour Grace Cable Station
Built in the 1830s by the Ridley Company, the building was converted to a cable office for the Direct United States Cable Company in 1910.

In 1873, AAT amalgamated with the New York, Newfoundland and London Telegraph Company and obtained the latter's monopoly on telegraph communications in Newfoundland. In the same year, AAT built a land line between Heart's Content and Placentia, which connected to a new submarine cable via St. Pierre and Nova Scotia. In 1875, AAT commissioned a permanent cable station building for Heart's Content, which was constructed by J. & J. Southcott of St. John's. AAT passed the original wood frame cable building over to the company employees, who used it as a recreation complex until it was demolished in 1955. Over time, with advances in technology, the Heart's Content station became obsolete and closed down on June 30, 1966, almost one hundred years after the first cable was landed there.

Harbour Grace (Direct United States Cable Company)

In 1874, the Direct United States Cable Company (Direct Cable) placed its first transatlantic cable. The cable ship *Faraday* laid the cable a few miles from the head of Trinity Bay, where it was attached to a buoy. Meanwhile, the company challenged AAT's monopoly on telegraphic communications in the Newfoundland courts. After losing its case, Direct Cable bypassed Newfoundland and terminated its cable at Tor Bay, about thirty miles south of Canso, in Nova Scotia.

Several decades later, AAT's monopoly expired, and in 1910, Direct Cable set up an office at Harbour Grace where it diverted its 1874 cable. By landing the cable in Newfoundland rather than Nova Scotia, the underwater length of the transatlantic section was shortened, resulting in a doubling of transmission speed. This allowed twice the number of messages to be carried.

The British government, under the name Imperial and International Communications Company, purchased Direct Cable in 1921. In 1935, the company changed its name to Cable and Wireless Limited. The Harbour Grace station, which was staffed by approximately a dozen employees, operated only one cable, and closed down in 1953.

St. John's (Commercial Cable Company)

In 1884, the Commercial Cable Company (Commercial Cable) installed two cables between Canso, Nova Scotia and Waterville, Ireland. It diverted the first of these in 1909 to Cuckold's Cove just north of St. John's, directly connecting Newfoundland with Canso and Ireland. The Cuckold's Cove office operated until 1916, when the company moved the station to a new location at 111 Water Street.

In 1905, the Newfoundland government implemented a $4,000 tax on submarine cable landings in the colony. Commercial Cable challenged this levy and refused to pay the tax. In 1917, the government gave notice that it would therefore terminate all contracts with the company within six months. The government ultimately won the court case and both parties eventually reached an agreement in 1922 allowing Commercial Cable to transmit and receive local telegraph messages to and from Newfoundland.

In 1926, Commercial Cable landed two transatlantic cables at Quidi Vidi Harbour. These cables, and also those landed in Cuckold's Cove, were buried to the east end of Quidi Vidi Lake. From there, they followed the bottom of the lake to the western end and were trenched to the company's Water Street office.

Commercial Cable Building, St. John's
This building is now the Brother Murphy Centre. (Courtesy of PANL)

After World War Two, the company employed about fifty operators. However, after years of dwindling business brought about by the introduction of transatlantic telephone service, Commercial Cable ceased its Newfoundland operation in 1961, affecting approximately forty employees.

Bay Roberts (Western Union Telegraph Company)

The American-owned Western Union Telegraph Company (Western Union) set up a telegraph station at Bay Roberts in 1910 to which it diverted its Coney Island, US to Penzance, UK cable. Two years later, it negotiated fifty-year operating leases on the cables owned by AAT and Direct Cable, including those at Heart's Content and Harbour Grace, giving it effective control of those companies while allowing them to operate under their own names. In 1913, Western Union installed a submarine cable between North Sydney and Colinet, St. Mary's Bay, where it connected to a trenched landline to the Bay Roberts station.

A wooden structure served as the first Bay Roberts cable station until the company built a large brick building on Water Street toward the end of World War One. The company also built a large staff house and attractive homes for its senior staff on Cable Avenue next to the building. In 1926, the company installed, via Bay Roberts, a cable between Hammel, USA and Penzance, UK, followed by two years later by another between Bay Roberts and Horta in the Azores. The Bay Roberts station employed more than thirty people at the height of its activity; however, by 1957, automatic repeating equipment was installed and the staff was reduced to nineteen. The Bay Roberts cable station closed down in 1969.

TELEPHONE COMMUNICATIONS
LOCAL TELEPHONE SERVICE

John Delaney, the government's postmaster, made Newfoundland's first telephone connection in March of 1878. His sets were built from a description in the March 31, 1877 edition of *Scientific American* magazine. The telephone link was set up between his home at 2 Monkstown Road, St. John's and that of John Higgins, a Post Office messenger at 48 Southwest Street (now Colonial Street). The circuit used a telegraph line that was normally used for Post Office business. There was no further evidence of telephone use in Newfoundland until 1884, when a telephone line was set up in St. John's between Archibald's Furniture Store at the junction of Duckworth Street and St. John's Lane, and the residence of its manager about half a kilometre away on Devon Row.

In 1885, AAT installed Newfoundland's first public telephone switchboard on the second floor of John Lindberg's jewellery store at 171 Water Street in St. John's. After it was destroyed in the great fire of 1892, the company built a new exchange at 276 New Gower Street above James Black's drygoods store. AAT sold its telephone system in 1899 to the Western Union Telegraph Company, which moved the exchange to a building east of City Terrace.

In the early to mid 1900s, many larger companies set up their own telephone exchanges. These included the Buchans Mining Company in Buchans and Millertown; the Anglo-Newfoundland Development Company in Grand Falls, Bishop's Falls and Botwood; Newfoundland Fluorspar in St. Lawrence; the Dominion Steel and Coal Corporation on Bell Island; and Bowater Pulp and Paper Mill in Corner Brook. The United Towns Electric Company also had an extensive telephone system with several telephone exchanges on the Avalon and Burin Peninsulas. The Twillingate Telephone and Electric Company, a private enterprise, was set up in 1913 by a group of citizens to provide telephone service to the town. Telephone service to other communities was the responsibility of Avalon Telephone Company in their franchise area, and the Department of Posts and Telegraphs in all other areas.

Department of Posts and Telegraphs

Although AAT held a monopoly on telegraph lines in Newfoundland, it was not particularly interested in providing telegraph service to the smaller towns and villages. To improve communications to these communities, the Department of Posts and Telegraphs (DPT) began to install lines on its own, and by the end of the nineteenth century had extended telegraph service to the major centres in the colony. The Reid Newfoundland Company also operated a telegraph system for its railway, but in 1901, transferred its telegraph assets to the government for an annual subsidy of $10,000, agreeing to operate the system until AAT's monopoly expired.

In 1902, the government installed new cable on the pole line between Whitbourne and Port aux Basques, and later extended it to St. John's. Two years later, it installed a submarine telegraph cable across the Cabot Strait from Port aux Basques to Canso, Nova Scotia where it connected into the Commercial Cable Company telegraph system. In the same year, the government set up Newfoundland Postal Telegraphs, a section of the DPT to consolidate its postal, telegraph and telephone operations.

In 1929, more than one million telegrams were handled by the government's telegraph service. More than ten thousand of these were sent by the Post Office as public service messages, such as messages regarding weather, shipping and fishing information. In the same year, DPT installed a small telephone exchange at Abraham's Cove on the Port au Port peninsula to serve the telephone subscribers in the area. By the end of 1930, DPT operated 246 telegraph offices and 170 telephone stations serving 347 communities. Fifty-five of the telegraph offices were wireless stations, eleven of which served com-

David Stott (1850-1929)
Stott, born in Aberdeenshire, Scotland, moved to Newfoundland in 1867 when he joined the New York, Newfoundland and London Telegraph Company, which later reorganized as Anglo-American Telegraph. In 1889, he entered the service of the Newfoundland Department of Posts and Telegraphs, becoming Superintendent in 1892.

munities and fishing stations on the coast of Labrador. DPT installed its first major telephone system in 1937 at the request of the British Air Ministry, to connect Gander airport with the Botwood seaplane base.

During World War Two, the United States Military built bases at Stephenville, Gander, Goose Bay, Argentia and St. John's. To connect these with one another and with bases in the United States, the Newfoundland Railway allowed the US Military to use its telegraph and telephone poles. The United States forces contracted with Bell Telephone Company of Canada, which sent more than three hundred of its own employees and hired more than seven hundred Newfoundlanders to work on the project. Construction began in April 1942 and was completed in less than a year later. From late 1942 to late 1945, the communications system was maintained by the US 821st Signal Wire Maintenance Company. Also during the war, the Canadian Armed Forces built a system over the government's pole line on the west coast, connecting to the Canadian mainland at Cape North, Cape Breton Island via a Very High Frequency radio system, installed at the top of Table Mountain.

In 1946, DPT took over the Royal Canadian Air Force (RCAF) manual telephone exchange in Gander and its pole system to St. John's. DPT also assumed the US Military's communications system to provide long distance telephone service to communities along the line. DPT operated their various communications networks until Newfoundland's Confederation with Canada in 1949.

Canadian National / Terra Nova Telecommunications

After Newfoundland joined Canada in 1949, responsibility for the telecommunications operations of the Department of Posts and Telegraphs was transferred to the Canadian government, which entrusted the assets associated with these services to Canadian National Railways (CNR). CNR's communications arm was Canadian National Telecommunications (CNT), which was given responsibility for operating the system. CNT inherited 932 telephone subscribers scattered throughout the remote areas of the province. J. H. Clarke, who was previously an employee with the DPT, became CNT's first superintendent in Newfoundland.

After taking over DPT's off-island submarine cable system, CNT installed a twelve-channel radio system across the Cabot Strait between Red Rocks (near Port aux Basques) and Sydney, Nova Scotia and also rebuilt the pole line to St. John's.

A magneto switchboard similar to those used in rural communities in the 1920s (Courtesy of Newtel)

In 1950, CNT purchased Western Union's operations in Newfoundland and assumed responsibility for all telegraph services in the province. The following year, it took over the telephone operations of the Twillingate Telephone and Electric Company and installed a new dial exchange. During the 1950s, CNT installed automatic dial telephone exchanges in several major towns and also purchased Buchans Mining Company's telephone system. During this period, CNT extended its telegraph network, providing service to locations such as Musgrave Harbour, Badger's Quay, Wesleyville, La Scie, Tilt Cove and Catalina.

Allan C. Jerrett became CNT's Newfoundland's superintendent in 1955. Two years later, the company consolidated its St. John's telegraph and telephone equipment as well as its administration offices into a new four storey building at 152 Water Street.

Communications engineering in the province took a giant leap forward in the late 1950s with CNT's design and construction of the province's first long-haul, heavy-route microwave system. This facility comprised five microwave channels, each capable of carrying six hundred telephone channels or one television channel. The project presented many engineering challenges. One was the design of the system itself, spanning several hundred miles of rough terrain not yet serviced by highway. The microwave facility required approximately twenty-seven diesel and radio buildings and forty-five diesel generators. The system was designed to meet stringent reliability requirements. The design of the microwave hop across the sixty-nine mile stretch of the Cabot Strait posed the most interesting engineering problems. CNT engineers selected the highest possible sites to provide a direct line of sight for the microwave signals. The normal distance between repeaters was usually less than forty miles. The engineers decided that extraordinary measures would be required to provide uninterrupted service in times of atmospheric and reflective disturbances. To achieve this, they employed both space and frequency diversity. In other words, four transmitters operating at different frequencies were used, located at different heights on the tower. At the receiving end, all four signals were received and logic circuits chose the best one. To allow for the long overwater span, the engineers also provided a much greater transmit power than that used for a normal microwave hop. At the time, the Cabot Strait section was one of the longest overwater microwave hops in the world

CNT completed the system between St. John's and Sydney, Nova Scotia, in 1959. In addition to providing high quality telephone circuits off the island, television signals between mainland Canada and Newfoundland were possible for the first time. Newfoundlanders could now view live news broadcasts and sporting events such as Hockey Night in Canada. The approximately six hundred mile long microwave system was by far the longest system built in the province to that point in time.

In 1967, Joseph Donich, a long term manager from CNT's mainland operations, succeeded Allan Jerrett. At the time, CNT operated eighty-two exchanges, of which sixty-seven had fewer than one hundred subscribers. In 1967, the company also completed a 260-mile long microwave system between Corner Brook and St. Anthony. CNT installed a spur between Mount St. Margaret and L'Anse au Loup in 1971, which linked into Bell Canada's telephone system in Labrador. In the same year, a three-hundred-circuit microwave facility was also installed between Corner Brook and Stephenville.

In 1979, CNR established Terra Nova Telecommunications Inc. (Terra Nova Tel) to take over Newfoundland's CNT operations. The company moved its headquarters to Gander in 1980, and Jack Gosse, who succeeded Joseph Donich in the late 1960s as CNT's superintendent in Newfoundland, became General Manager of the new company. Several months later, Gosse retired and was succeeded by Robert Symonds.

In the 1980s, digital switching technology became available and began to replace analog equipment. Terra Nova Tel installed its largest and most significant digital switch at Gander in 1985, which replaced the electromechanical system that had served the town for twenty years. In 1987, Terra Nova Tel completed a five-year improvement program, which provided the company's subscribers with 100 per cent Direct Distance Dialling and 99 per cent single party service. At the time, Terra Nova Tel had more than fifty thousand subscribers in more than four hundred communities throughout the province. In 1988, CNR decided to divest itself of its telecommunications assets and put Terra Nova Tel up for sale. On December 1, 1988, Newfoundland Telephone became the new owner and began merge the two operations.

Avalon / Newfoundland Telephone Company

John J. Murphy, along with his son Robert J. Murphy and J. D. Cameron, formed the Avalon Telephone Company (Avalon) in 1919. Shortly after setting up, Avalon purchased Western Union's telephone system in St. John's, which at the time served approximately eight hun-

John Joseph Murphy (1849-1938)
Murphy, who was born in St. John's, began his career with Ridley and Sons and was made manager of their Greenspond operations. Murphy purchased the concern after the Ridleys went bankrupt. His career spanned many areas, including lumbering, electric utilities, communications as well as politics. He opened a lumber mill near Gambo in 1876, which he sold in 1904, and decided to enter the emerging electrical business. He invested in United Towns, eventually becoming major shareholder and president in 1915. He also expanded into the communications business and in 1919 founded Avalon Telephone Company Limited, becoming president – a post he held, along with his position with United Towns, until his passing in 1938. (Courtesy of Newtel)

dred telephone subscribers from an exchange on New Gower Street. In 1921, the company installed a seven thousand line telephone exchange at 348 Duckworth Street in St. John's. At the time, telephone rates were $30 a year for a residence phone and $40 a year for a business phone. In the same year, the company also installed its first long distance line between St. John's and Carbonear, and on November 27 the first long distance call was placed from St. John's to Brigus and Harbour Grace. Also in 1921, the company constructed a telephone line along the Southern Shore to Cape Race, which enabled telephone service to be provided to communities along the route. In 1925, Avalon expanded service to Bell Island when it placed a submarine telephone cable under Conception Bay. By 1930, Avalon Telephone had about six thousand subscribers, with telephone exchanges in most of the towns in the Conception and Trinity Bay areas.

In 1938, John Murphy passed away and was succeeded as president by his son Robert, an electrical engineering graduate from the Massachusetts Institute of Technology. During John Murphy's presidency, all telephone communications used land lines; Robert Murphy's tenure heralded the introduction of radio facilities for telephone communications. In 1939, the company installed High Frequency (HF) radio facilities to provide telephone circuits from St. John's to the Burin Peninsula, Grand Falls and Corner Brook areas. Avalon inaugurated long distance service between Newfoundland and Canada on January 10, 1939, with a call from St. John's to Ottawa between Sir Humphrey Walwyn, Governor of Newfoundland and Lord Tweedsmuir, Governor General of Canada. The circuit was carried by an HF radio system owned by Canadian Marconi which remained the only telephone link between Newfoundland and Canada until Confederation in 1949. Following the inaugural ceremony, the first overseas telephone call between St. John's and London, England was also made. At the time, the rate for a three minute call from St. John's to Montreal was $7.50 and to London $22.20. Comparing these rates with those in 2002, long-distance competition has forced rates to as low as about three cents per minute between North America and Australia.

Robert J. Murphy (1891-1980)
R. J. Murphy, the son of John Joseph Murphy, was president of Avalon Telephone and United Towns between 1938 and 1954. He graduated from the Massachusetts Institute of Technology with a Bachelor of Science in Electrical Engineering. Murphy was one of the first Newfoundland-born engineers to oversee a major company. (Courtesy of Newtel)

In 1944, Avalon took over Bowater Pulp and Paper Mill's twenty-five-year-old telephone system in Corner Brook, and in 1947 replaced it with a new automatic exchange. In 1948, it opened a new five storey administration centre at 343 Duckworth Street in St. John's and introduced Newfoundland's first dial service in the capital city.

Avalon established telephone service between St. John's and Port aux Basques in 1949, allowing communities along the line to connect to others from one end of the island to the other. In 1951, the company acquired Anglo-Newfoundland Development Corporation's telephone exchanges in Grand Falls, Botwood and Bishop's Falls. By the end of 1952, Avalon Telephone had 23,509 telephones in service, of which

16,217 were in St. John's, 2,890 in other communities on the Avalon peninsula, and 4,402 on the west coast and Grand Falls. In 1954, Avalon changed hands when a group of Newfoundland and Montreal investors purchased control of the company. Robert Murphy resigned and Sydney H. Morris became president.

On December 16, 1955, Mayor Harry Mews of St. John's opened a ten thousand line dial exchange at a new building on Anderson Avenue. In 1957, Avalon Telephone became a member of the Trans-Canada Telephone System, a consortium of major Canadian telephone companies that co-ordinated long distance telephone traffic between member companies. Avalon Telephone installed its first microwave system in 1958, linking St. John's with Bell Island and Bay Roberts, via a repeater station on Kenmount Hill.

The company added twelve hundred telephones to its network when it purchased United Towns' telephone operations on the Burin Peninsula in 1962. In the same year, Avalon Telephone was purchased by the Bell Canada Telephone Company, but continued to operate under its own name. Avalon appointed George C. Wallace, a Bell Telephone senior manager, as President and Managing Director. He was succeeded two years later by Gunder Osberg, another Bell Canada employee.

In 1966, the company installed a state-of-the-art twenty-thousand-line crossbar switching machine in its new Allandale Road building in St. John's. Coincident with this installation, push button telephone service became available to St. John's subscribers.

On January 1, 1970, Avalon changed its name to Newfoundland Telephone Company Limited (Newfoundland Telephone). In the same year, Osberg returned to Bell Canada and was replaced by Anthony A. Brait, P. Eng., who had earlier worked with Avalon as Chief Engineer. Also in 1970, the company began its Direct Distance Dialling (DDD) program, which allowed subscribers to call long distance without having to go through an operator. The following year, an era ended as the company retired its last magneto offices at Western Bay and Old Perlican. The old crank style magneto phones immediately became collector's items.

At Corner Brook, the company installed its first computer-controlled stored program electronic switch in 1974, followed by similar installations in St. John's, Mount Pearl and Grand Falls. Also in that year, Newfoundland Telephone purchased Bell Canada's assets in Labrador, and in 1975 installed a microwave system between Goose Bay, Hopedale and Nain. This system provided many engineering challenges, such as the design of towers and buildings that would withstand the heavy icing conditions in the area. The engineers also introduced no-break diesel generation to power the sites. Providing these remote sites with fuel and performing general maintenance was a difficult proposition during the summer season, not to mention the winter months.

In 1977, the company installed a light route microwave system between L'Anse au Loup and Charlottetown, Labrador providing long distance telephone service to the latter community. In the same year, it also installed a four-hundred-mile-long microwave system between Goose Bay and L'Anse au Loup which connected to CNT's microwave system to Corner Brook, where it accessed Newfoundland Telephone's network.

Anthony A. Brait, O.C., P. Eng. (1924-1996)
Anthony (Tony) Brait became Chief Engineer of Avalon Telephone in 1964, and was appointed President in 1970. During his tenure, he led the company into the age of digital technology and expanded the company to cover the entire province. He received the order of Canada in 1993. (Courtesy of Newtel)

In 1978, Newfoundland Telephone completed its first trans-island microwave system, extending from St. John's, via Grand Falls and Corner Brook, to Sydney, Nova Scotia. The facility provided twelve hundred voice circuits and also allowed the company to carry television signals into and out of the province. This system – as had the one built by CNT almost twenty years earlier – presented many engineering challenges, particularly the long overwater hop across the Cabot Strait.

Also in 1978, Newfoundland Telephone entered into an agreement with Eastern Telephone and Telegraph Company (ET&T), a subsidiary of American Telephone and Telegraph Company (AT&T), for the purchase of its assets in Clarenville. ET&T's main function was the man-

> **Shoe Cove Satellite Tracking Station**
> *In 1971 the American National Aeronautical and Space Administration was in the middle of its Apollo program and concerned about a communications blind-spot over the North Atlantic. They subsequently obtained permission from the Canadian government to construct a satellite tracking station at Shoe Cove, just north of St. John's, to fill this gap. The station opened in 1971 and closed in 1977 after new technology obviated its need. At its peak there were sixty U.S. technicians at the site as well as seventeen Newfoundlanders.*

agement of AT&T's transatlantic telephone cables which came ashore in Clarenville. Newfoundland Telephone continued this responsibility until the last of the cables went out of service in 1982.

In 1979, Newfoundland Telephone acquired the Iron Ore Company of Canada's telephone assets in Labrador City, adding sixty-five hundred telephones to its network. That year also marked a new era in communication engineering – the introduction of digital technology. The company's first digital switch was installed in 1979 in Nain, Labrador, followed in 1980 by Corner Brook. Just about all future exchanges were to employ digital technology. With the older analog technology, communications links were either hard-wired or switched by electromechanical devices. The emergence of digital technology allowed engineers to configure their systems for the first time using software control.

In 1981, the company moved to a new administration building on Factory Lane in St. John's. The following year Telesat Canada built a satellite earth station on Kenmount Road in St. John's. The satellite station connected to Newfoundland Telephone's centre at Allandale Road via its first optical fibre link and was part of a Canada-wide video conferencing system. Some years earlier, in 1970, Telesat Canada had built a transmit and receive satellite station at Bay Bulls; in 1975 it also constructed one on the Port au Port peninsula to receive French language television.

After Newfoundland Telephone completed its trans-Labrador microwave system to Wabush/Labrador City in 1983, it decommissioned its last troposcatter system in Labrador. The troposcatter systems used 60 to 120 foot high antennas to bounce radio signals off the troposphere, high above the earth's surface. This system was built by the United States Armed Forces in the early 1950s as part of the ballistic missile early warning system.

A microwave site in Northern Labrador, illustrating the extreme environmental conditions which had to be taken into account in designing these systems (Courtesy of Newtel)

On October 15, 1985, in a corporate re-structuring, Newfoundland Telephone became a subsidiary of Newtel Enterprises Limited, a wholly owned subsidiary of Bell Canada Enterprises. In June 1986, it completed a $23 million digital microwave system between St. John's and Port aux Basques, and extended it from Corner Brook to St. Anthony in 1990.

As mentioned earlier, Newfoundland Telephone acquired Terra Nova Tel in 1988. This made Newfoundland Telephone the only company providing telephone service in the province. In 1989, Anthony Brait retired, and Vincent G. Withers, a long time manager with the company, became President and Chief Executive Officer.

Newfoundland received its first cellular telephone service in July 1990, when both Newfoundland Telephone and Rogers Cantel Communications set up networks in the St. John's area. Both companies expanded their networks to cover the Trans-Canada Highway to Clarenville, while Newfoundland Telephone also enlarged its network to cover the entire Trans Canada Highway across the province as well as to Labrador, the Burin peninsula and other major centres. In 1993, Newfoundland Telephone transferred its cellular division to Newtel Mobility, a wholly owned subsidiary of Newtel Enterprises.

Newfoundland Telephone began the installation of a $57 million optical fibre cable system between Sydney and St. John's in 1991. The submarine portion across the Cabot Strait and the underground portion to Corner Brook were completed later in the year. The cable was extended to Clarenville in 1992 and to St.

The Cape Race Loran C tower was 1350 feet high, one of the tallest such towers in the world. It collapsed in February, 1993, after almost thirty years of service. (Courtesy of the Canadian Coast Guard)

John's in 1993. In 1995, the company constructed a second optical fibre cable across the province with an underwater link between Codroy and Dingwall, Nova Scotia. This link was another engineering challenge because of the long underwater section. At the time, the submarine portion, at 180 kilometres, was the longest stretch of unrepeatered optical cable in the world.

In 1998, Newfoundland Telephone merged with the other three Atlantic province telephone companies to form Aliant Inc., with headquarters in Charlottetown, Prince Edward Island.

Unitel Newfoundland

After Confederation with Canada in 1949, the Newfoundland government set up the Board of Commissioners of Public Utilities (PUB) to regulate rate changes and other operations of the provincially incorporated telephone companies. Forty years later, in August 1989, a decision of the Supreme Court of Canada brought Newfoundland Telephone and several other Canadian telephone companies under the regulatory jurisdiction of the Canadian Radio-television Telecommunications Commission (CRTC), relieving the PUB of the responsibility of regulating provincial telecommunications.

After public hearings in June 1992, the CRTC determined that other carriers could compete with Canadian telephone companies for the public long distance business. In early 1991, Unitel Communications Inc. (Unitel), one of the parties which had initiated the proceeding leading to that decision, reached an arrangement with Fortis Properties Corporation to provide communications services in the province. These companies set up Unitel Newfoundland which constructed a digital radio facility from St. John's to Nova Scotia, connecting into Unitel's national microwave system at Cape Breton Island. Unitel, which was later acquired by AT&T Canada, began to provide long distance telephone service in Newfoundland in July 1993.

TRANSATLANTIC VOICE COMMUNICATIONS
St. John's (Canadian Overseas Telecommunications Corporation)

Newfoundland's role in transatlantic communications systems was not limited to telegraph systems. In 1950, the Federal government established the Canadian Overseas Telecommunications Corporation

(COTC) to acquire the transatlantic communications assets of Cable and Wireless Limited and the Canadian Marconi Company. COTC changed its name to Teleglobe Canada in 1975. COTC participated in the installation of the first transatlantic telephone cable, which began on June 22, 1955 when the *Monarch* left Clarenville for Oban, Scotland paying out the "TAT-1" cable. The $42 million venture was financed by AT&T, the British Post Office and COTC. TAT-1, with a capacity of thirty-six voice circuits, extended overland from Clarenville to Terrenceville on the Burin peninsula and then by submarine cable to Sydney Mines, Nova Scotia. The cable remained in service until 1978. In 1959, COTC and partners installed the TAT-2 transatlantic telephone cable, providing an additional forty-eight transatlantic telephone circuits. Eastern Telephone and Telegraph Company originally managed the Clarenville cable station but later sub-contracted the work to Newfoundland Telephone. The station remained in service until TAT-2's retirement in 1982.

In 1961, COTC participated in the installation of the eighty-circuit CANTAT-1 cable, between Oban, Scotland and Hampden, White Bay; the cable connected via Deer Lake to Wild Cove, near Corner Brook, and then via submarine cable to Grosse Roche, Quebec.

In 1974, CANTAT-2, a 1.47 inch co-ax cable carrying up to 1840 voice circuits, was installed between Widemouth, England and Beaver Harbour, Nova Scotia. The vastly increased circuit capacity of this cable rendered CANTAT-1 obsolete and it was taken out of service. As a result, COTC's Deer Lake cable station closed down in 1975, followed shortly thereafter by its Wild Cove location.

OTHER COMMUNICATIONS OPERATIONS
MICROWAVE SYSTEMS

Newtel and AT&T are not the only companies with communications facilities in the province. Many concerns – such as the federal and provincial departments of communications, paging companies, and broadcasting companies – also have their own communications links. The most significant communication systems apart from those of the common carriers are those owned and operated by Newfoundland and Labrador Hydro and Newfoundland Power to monitor and control their generating and distribution plants.

ADMIRALTY HOUSE

In 1915, the British Admiralty constructed a $500,000 wireless station at Mount Pearl to provide weather reports and coded messages to the British Navy during World War One. The station's code was BZM and its transmitter was powerful enough to reach Bermuda and the eastern part of the United States. Occasionally the station could communicate with locations as far away as Ireland and England. The station had three 305 foot high steel towers supporting the antenna and a thirty-kilowatt Marconi synchronous spark transmitter. The station remained in service for nine years, after which it was put up for sale, but there were no takers. The Admiralty House museum, in Mount Pearl, now occupies the site of the station and houses a number of communications artifacts.

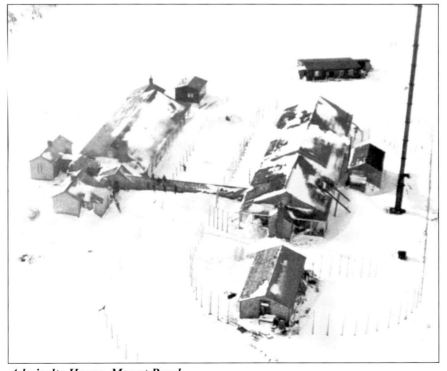

Admiralty House, Mount Pearl
This facility was constructed by the British Admiralty in 1915 to monitor shipping in the North Atlantic (Courtesy of Admiralty House Museum)

CAPE RACE LORAN STATION

The Canadian Coast Guard has a number of sites around the island to aid marine transportation. One of these is at Cape Race where a LORAN C (Long Range Aid to Navigation) station employs a 850-foot-high tower to beam powerful radio signals to ships at sea. The station had previously used a 1350-foot-high corner guyed steel tower which collapsed in 1993. At the time it was one of the world's highest guyed towers.

WIRELESS COMMUNICATIONS
NEWFOUNDLAND'S EARLY WIRELESS

Newfoundland first experienced wireless transmission on September 12, 1899. On that date, a Mr. Bowden displayed wireless telegraphy in St. John's. A. M. Devine, editor of the *Trade Review* newspaper, was asked to write a message to test the system and responded with "God save the Queen."

Prior to Marconi's radio experiment in 1901, the Canadian government had established Marconi stations at Belle Isle and Chateau Bay to communicate with shipping in the area. After the turn of the century, Marconi stations were also constructed at Cape Race, Cape Ray, and Point Riche. By 1906, the government had established wireless stations at several small fishing stations in Labrador, including Battle Harbour, Venison Island, American Tickle, Domino and Indian Harbour. After 1910, merchants also began to outfit their sealing ships with Marconi radio. Over the years the government relied heavily on wireless communications, and up until the mid 1900s, many of Newfoundland's remote communities were still served with Marconi wireless systems.

MARCONI AND THE WIRELESS TELEGRAPH

Almost two years after wireless was first demonstrated in Newfoundland, the *SS Sardinian* sailed into St. John's harbour. On board was Guglielmo Marconi, the famous Italian-English inventor, along with his associates, George Kemp and Percy Paget. They arrived in St. John's on December 6, 1901 without fanfare. Marconi had not previously announced his plan to attempt a reception of a transatlantic wireless signal, probably concerned that a failed experiment would harm the reputation of his company. To disguise his true intentions, he told reporters and government officials that he was in Newfoundland to investigate the possibility of setting up a wireless station to communicate with shipping in the area. Officials arranged for Marconi to use space in a former hospital on Signal Hill for his experiments. The hospital was close to Cabot Tower, which had recently been built to commemorate the four hundredth anniversary of John Cabot's landing in Newfoundland and the sixtieth anniversary of Queen Victoria's reign.

On Monday, December 9, Marconi began installing his apparatus and by Thursday, December 12, after earlier testing the kites and losing a

Guglielmo Marconi
Marconi at his wireless reception apparatus on Signal Hill in 1901 (Courtesy of PANL)

balloon in a gale, he was ready to begin his experiment. His group successfully launched a kite hoisting the aerial, and Marconi and Kemp began monitoring the receiver inside the old hospital, leaving Paget to manage the nine-foot-long kite in a strong wind. Around noon, at the time that he had instructed the wireless station at Poldhu, Cornwall to start transmitting, Marconi listened into the telephone receiver that was connected to his receiving apparatus. Shortly afterward he heard three faint clicks in the receiver designating in Morse Code the letter "s." He waited for the signal to be repeated and again he heard the three dots transmitted from almost two thousand miles away. The kite dropped for a short time, but lifted off again, allowing Marconi to faintly hear the transmission again. Marconi's great experiment had succeeded!

Marconi did not want to announce his successful experiment to the newspapers until he could repeat it to satisfy himself that the result was genuine. The next day, he and his team again went to Signal Hill, sent up the kite, and once more heard the transmission from Cornwall. He ensuingly announced his transatlantic reception to the local press which quickly spread the news via cable to all corners of the world.

Marconi already had plans for a commercial transatlantic wireless system. He recognized the economic advantages of wireless telegraphy, as its setup cost was only a fraction of that of an underwater cable. The stock markets were quick to realize this and reacted sharply. Almost overnight, the price of transatlantic cable company shares plummeted.

For the North American terminus of his wireless system, Marconi considered a number of locations, including Cape Spear, the most easterly point in North America, and on December 16 visited the site to assess

Launching a kite
Marconi and his helpers preparing a kite for launch on Signal Hill in December 1901 (Courtesy of PANL)

its suitability. On his return a representative of the Anglo-American Telegraph Company was waiting for him at the Cochrane Hotel. AAT, it should be recalled, had a fifty-year monopoly on telegraphic communications in Newfoundland. Although only three years were remaining on its monopoly, the company was firm in protecting its rights. The representative presented Marconi with a letter threatening court action if he continued his wireless experiments. A. M. Mackay, AAT's local manager, who almost fifty years before had headed the repair of the St. John's to Cape Ray telegraph cable, led the fight to keep Marconi out of Newfoundland.

Not wanting court action to delay his planned transatlantic system, Marconi abandoned any plans to build a terminus in Newfoundland, and responded to an invitation from the Canadian Government to set up in Nova Scotia. On December 24, he took the train to Port aux Basques to connect with the gulf ferry. At the St. John's Fort William train station, near the site of the present Hotel Fairmont and Fort William Buildings, he was sent off by a large and enthusiastic crowd of well-wishers.

Marconi's departure caused great concern in some quarters and many even doubted the validity of the monopoly claimed by Anglo-American Telegraph. Despite the protestations against AAT's monopoly, Marconi's relationship with Newfoundland had ended. He had been in St. John's for only eighteen days, but during that time, had broken through an engineering barrier, and had proven that transatlantic wireless communication was possible. His exploits had been reported and praised around the world. The use of wireless had been firmly established and the demise of telegraph cable was only a matter of time.

BROADCASTING
RADIO

In its early days, "wireless" had been used only for telegraphic communications. That changed in 1903 when Reginald Fessenden (1866-1932), demonstrated that voice communications could also be transmitted using wireless. Several years later, in 1906, from Brant Rock, Massachusetts, he broadcast his voice as well as a violin rendition of "O Holy Night" over the airwaves; this was received by several ships hundreds of miles out to sea. Fourteen years later, in 1920, station KDKA became the world's first radio broadcast station when it went on the air in Pittsburgh.

In the same year, three engineers from Canadian Marconi arrived in St. John's and set up a wireless station for voice communications with a ship sailing from the other side of the Atlantic. Overseeing the project was J. J. Collins, Canadian Marconi's representative in Newfoundland and one of Newfoundland's early radio pioneers. On July 23, Collins and his group made contact with the *SS Victorian* shortly after it had left England. The ship at the time was en route to a newspaper convention in Quebec City with a large contingent of press people. The radio contact was the first between North America and a ship on the other side of the Atlantic. The signal was strong and communication was maintained for the duration of the voyage. Two days later, local dignitaries came to Signal Hill to participate in this engineering first, including prime minister Sir Richard Squires and St. John's mayor W. G. Gosling, both of whom exchanged words with the crew and guests of the *Victorian* while it steamed toward Quebec City.

The first Canadian radio broadcasting station was XWA (now CFCF) in Montreal, which began regular programming on May 21, 1921. Only a year later, Collins set up Newfoundland's first broadcasting operation. The station, with code letters VOS, was located at the Pope building on McBride's Hill in St. John's. It operated for only two years, but for those with a radio receiver, was the first taste of what was to become a way of life. The station's call letters were changed to 8AK in 1923 and then to 8JJC (using his initials) after Collins moved the equipment to his home on Parade Street. Collins also set a Newfoundland first when he made the first remote radio broadcast – a live coverage of a hockey game on January 25, 1925 from the Prince's rink in St. John's. The broadcast was sent over telephone lines to his studio, where it was transmitted to an appreciative audience.

Reverend Joseph G. Joyce (1889-1959)
A pioneer in Newfoundland radio, Joyce started up station VOWR in 1924. (Courtesy of VOWR)

By the early 1920s, both crystal and tube radios were easily available, and early Newfoundland radio listeners could use their sets to receive radio signals from as far away as the United States. Local listeners were given another option when Ern Ash set up Station 8AA in 1923. The following year, Reverend J. G. Joyce (minister of Wesley Methodist Church between 1922 and 1930) foresaw his parishioners' desire to have church services broadcast, especially for the aged, infirm, and isolated who could not get to church. Initially, he used telephone lines to carry sermons to home-bound parishioners. This was not the first time a church service had been transmitted to parishioners over telephone lines. As early as 1882, Reverend C. Smith in Heart's Content used a telephone connection to transmit his Sunday sermons to the bedridden H.C. Weedon, superintendent of AAT's cable station.

Joyce realized that transmitting church services via telephone was not practical, so he turned to radio. He obtained a small one hundred watt transmitter from Canadian Marconi and installed it at his church on Patrick Street. The station, which was designated 8WMC, started operations on July 20, 1924 and initially operated two evenings a week, broadcasting church programs, weather forecasts, classical music, and talks on subjects of current interest. The Boy Scouts helped generate interest in the radio station by making and distributing inexpensive crystal radio sets to those that needed them.

In 1932, 8WMC was changed to VOWR (Voice of Wesley Radio), by which it is still known today. The station's power was increased to one thousand watts in 1954 and five thousand watts in 1984.

In the early 1920s, several other radio stations went into service. On November 4, 1924, the Reid Newfoundland Company set up station 8LR, which on its opening day played gramophone records of Gilbert and Sullivan's opera "The Gondoliers." A year later, William F. Galgay helped the Mount Cashel Christian Brothers establish 8MC. In 1929, Ern Ash set up a short wave broadcasting station from his residence on Patrick Street. In the same year the Department of Posts and Telegraphs granted operating licenses to fourteen amateur radio experimental stations.

Another church-operated station went on the air in 1929, when the Seventh Day Adventist Church, under pastor Reverend Harold N. Williams, set up station 8BSL (Bible Study League), which was dedicated to broadcasting church and religious programming.

The 1930s saw the establishment of a number of new radio stations. In 1931, Ayre and Sons, a large St. John's department store, wanted to promote the sale of its radio sets. Working with the company at the time was nineteen year old Oscar Hierlihy, who was substituting for his sick brother. Hierlihy in his youth had experimented with radio, and had even transmitted gramophone records from his home at Bay Roberts to radio sets in the immediate vicinity. Ayre and Sons offered him a job in their radio department, and asked him to build a transmitter to demonstrate the radios they had for sale. Along with his brother Cliff and W. F. Galgay, Hierlihy set up radio station VOAS or Voice of Ayre and Sons, as part of the firm's advertising department. Originally the station was only used to demonstrate the store's radio sets, but Hierlihy later erected an outdoor antenna allowing the station to be received as far away as Bay Roberts. This was Newfoundland's first radio station used for advertising purposes. Hierlihy stayed with Ayre and Sons for a year, after which the company closed VOAS down because it had no one to operate it.

> **Broadcast Church Services**
>
> The broadcasting of church services was still such a novelty that on 4 May 1929 the *Toronto Star Weekly* published the following:
>
> <u>Little Newfoundland Church provides religion by air.</u>
> *When the minister of an unpretentious little United Church in St. John's, Nfld., steps into his pulpit to begin his Sunday Service, a great unseen audience many of them marooned far from the footsteps of the world's traffic, waits expectantly.*

When the pastor of the Seventh Day Adventist Church was transferred to the United States, no one knew how to operate its radio station (8BSL), so he handed it over to Oscar Hierlihy on the condition that he broadcast the Sunday church services. Hierlihy took over the station, changed its call letters to VONA (Voice of the North Atlantic), and for about a year operated it as Newfoundland's first commercial station. In 1936, the church again put its own station on the air broadcasting religious programming as VOAC (Voice of Adventist Church). The station's designation changed again in 1938 to VOAR, its current call letters.

In 1931, the International Grenfell Association set up VOR at St. Anthony to serve the northern part of the island and Labrador. In May of the same year, Rev. E. J. Rawlins established station VOGT (Voice of Gaiety Theatre) on Bell Island. During the same year, C.L. Parkins of the Imperial Manufacturing Company aired VOKW, which was used primarily for advertising purposes. This station also shared broadcast time with the Royal Stores Ltd.

In 1932, Frank Wood formed the Newfoundland Broadcasting Company and established VOGY, which also incorporated the facilities of VOLT which was set up by Ern Ash and Bill Wood earlier that year. VOGY's studio was located at the Crosbie Hotel on Duckworth Street

Oscar Hierlihy, CM (1912-1996)
One of Newfoundland's pioneers in radio and television broadcasting, Hierlihy was instrumental in setting up several early radio stations and CJON TV. He became a member of the Order of Canada in 1991. (Courtesy of PANL)

and its transmitter was on Mundy Pond Road. VOGY, which broadcast its first program on September 12, became Newfoundland's first full-time commercial station. In March 1934, the station moved its studios to the Newfoundland Hotel.

Also in 1932, the Dominion Broadcasting Company, a subsidiary of Avalon Telephone, established VONF. The station, which was situated on McBride's Hill, was set up by Oscar Hierlihy and had a power of five thousand watts. By this time Hierlihy had closed down both VONA and VOAS, and had become an employee of Avalon Telephone. William Galgay, who had previously worked with Hierlihy at Ayre and Sons' radio station, was VONF's studio director. Joseph L. Butler, who would later found VOCM, was also on the staff. Besides broadcasting on AM, VONF also broadcast as shortwave station VONG, which gave the station a much greater range.

The government set up the Broadcasting Corporation of Newfoundland in 1938, which took over the Dominion Broadcasting Company the following year. William Galgay served as General Manager until the Canadian Broadcasting Corporation (CBC) acquired the company in 1949, at which time he became head of the CBC in Newfoundland.

One of the more popular programs aired by VONF was the Gerald S. Doyle Bulletin, which was primarily meant to serve the isolated outports scattered around the coast of colony by broadcasting personal messages, weather forecasts, shipping reports and other local news. In 1932, VONF aired the first election report and in 1934, broadcast the St. John's regatta. VONF and VOGY merged in 1934, but continued to operate as VONF. The station was transferred to the Canadian Broadcasting system after Newfoundland joined Canada in 1949, and was renamed CBN. It moved to the TA (Temperance and Abstinence) building on Duckworth Street, where it still operates today.

In 1936, Joseph L. Butler and Walter B. Williams introduced VOCM, a 250 watt radio station with a studio at the corner of Parade Street and LeMarchant Road. Later the station moved to the top floor of the Pope building on McBride's Hill, where it remained until 1967, when it moved to a new building on Kenmount Road. The Colonial Broadcasting System, which owned the station, expanded over the years and added radio stations at Marystown (CHCM) in 1961, Grand Falls (CKCM), and Gander (CKGA) in 1968, Baie Verte (CKIM) in 1971, Clarenville (CKVO) in 1974 and Carbonear (CHVO) in 1980. Joseph V. Butler, the son of the founder, became the president of the Colonial Broadcasting System in 1961.

The first radio station in Labrador was established by Reverend Lester L. Burry in 1937. Burry, a United Church missionary, would become one of Newfoundland's fathers of Confederation in 1947 when he served as a member of Newfoundland's delegation to Ottawa. He set up a small transmitter at his house at North West River, Labrador, which he used to make broadcasts to the local area. In 1938, the government issued the station (VO6B) a commercial license to transmit church broadcasts, news and general information to the area around Hamilton Inlet.

The radio career of one of Newfoundland's most illustrious sons began on station VONF. In 1937, Joseph R. Smallwood began broadcasting "The Barrelman" program, in which he regaled the listening audience with stories of Newfoundland and Labrador. This program made him a household name and undoubtedly contributed to his later political success. Smallwood led Newfoundland into Confederation with Canada in 1949 and became the province's first premier.

The 1940s likewise provided an eventful decade in Newfoundland's radio history. Stations were established to serve the large contingent of military personnel that was stationed in Newfoundland during World War Two. In 1941, radio station VONF broadcast the arrival of the *Edmund B. Alexander*, the largest ship to have entered St. John's harbour up to that time, which was bringing the first contingent of American Armed Forces personnel to Newfoundland. At Pepperrell Air Force Base in St. John's, radio station VOUS (Voice of United States) started up on November 1, 1943. This station broadcast material of interest to the American troops, until the base closed down, after which it was transferred to the US Naval Base at Argentia. On December 11, 1943, the RCAF set up VOUG in Goose Bay, followed on January 1, 1944 by VORG (Voice of Radio Gander) in Gander to serve the residents and staff at these air force bases. VORG's first studio was located in the basement of the Commanding Officer's home.

In 1943, the west coast of the island saw its first radio station when VOWN (Voice of Western Newfoundland) came on the air broadcasting from a studio on the top floor of the Glynmill Inn in Corner

Joseph (Joey) R. Smallwood (1900-1991)
Premier of Newfoundland (1949-1972), he became known to the majority of Newfoundlanders with "the Barrelman" radio program during the late 1930s. (Courtesy of MUN CNS Archives)

Brook. The station's staff consisted of Clifford Hierlihy and announcer John Grace.

On April 1, 1949, the Canadian Broadcasting Corporation took over all stations operated by the Broadcasting Corporation of Newfoundland and changed the station designations to conform with CBC regulations. VONF in St. John's became CBN, VOWN in Corner Brook became CBY, and VORG in Gander became CBG. CBT in Grand Falls, which was being installed when Newfoundland joined Canada, went on the air July 1, 1949. Privately owned radio stations in Newfoundland were allowed to retain the "VO" call letters. In 2002, only three stations remained with the "VO" designation – VOCM, VOAR and VOWR.

In 1950, Geoff Stirling and Don Jamieson formed the Newfoundland Broadcasting Company (NBC) – not to be confused with the company of the same name formed in 1932 – and launched radio station CJON in St. John's. Its studio was at Buckmaster's Circle and its transmitter four miles away on Groves Road. The station was designed by Oscar Hierlihy who, although still an employee of Avalon Telephone, worked with CJON in his spare time. On October 11, 1951, the station went on the air with Jamieson as its first announcer. In 1964, NBC opened station CJCN in Grand Falls and CHOZ in Grand Bank. In the mid 1970s, NBC closed down CJCN and CJOZ and decided to sell its AM stations to the newly formed Newfoundland Q Radio System. After deciding to get out of the AM radio business, the Newfoundland Broadcasting Company established OZ-FM, a network of nine FM stations around the province.

By the early 1950s, approximately 50,000 radio receiving sets served Newfoundland's 350,000 residents scattered over approximately thirteen hundred communities. Most communities could receive a radio station, but in many locations, reception was sporadic. The CBC, over the next few years, embarked on a program to ensure that all Newfoundland communities had reasonable reception of at least one station. This was done by using inexpensive low power relay transmitters which broadcast only to the immediate area in which they were located. These transmitters re-broadcast radio programming originating from CBC's regional radio stations, which was delivered to the transmitters over telephone lines.

In 1960, Dr. Noel Murphy organized the privately owned Humber Valley Broadcasting Company, which on October 5 launched station CFCB in Corner Brook. Radio broadcasting from CFCB was retransmitted by stations CFSX in Stephenville (1964), CFNW in Port aux Choix (1972), CFDL in Deer Lake (1974), and CFNN in St. Anthony

(1974). CFGN in Port aux Basques was set up in 1971, with a satellite transmitter at CFCV in St. Andrews (1974). In 1971, the company expanded to Labrador, opening AM stations CFLN in Goose Bay (1971), with satellite transmitters at CFLW in Wabush (1971) and FM station CFLC in Churchill Falls in 1974.

The Q Radio System was established on September 1, 1977 by Don Jamieson and was later sold to CHUM Limited of Toronto. Q Radio reopened the former CJCN in Grand Falls as station CIYQ, and CJOZ in Grand Bank as CKYQ. In 1983 CHUM sold its stations to NewCap Broadcasting Limited.

The CBC introduced FM broadcasting to Newfoundland in 1975 with station CBN-FM in St. John's. During this year, MUN Radio, the student radio station at Memorial University in St. John's, was granted a carrier current license to transmit student programming throughout the campus; ten years later, the CRTC granted it a low-powered FM broadcast license allowing it to go on the air as CHMR. In 1980, the CBC also started a French language FM station in St. John's, broadcasting its feed from Moncton, New Brunswick. On June 15, 1977, CHOZ FM went on the air, followed by VOCM FM in 1982 and CKIX in 1983. In 1988, VOWR applied to the CRTC for a license to operate an FM station but was denied on the grounds that it did not license religious undertakings.

As of 2002, apart from the church-operated stations VOWR and VOAR, Newfoundland Capital Corporation owned fifteen of the province's sixteen privately owned radio stations while the Newfoundland Broadcasting Company owned the other. The remainder

Donald Jamieson (1921-1986)
An active politician in his later years, most prominently as a Federal Cabinet Minister during the 1960s, Jamieson began his career as a founder of and broadcaster with CJON TV. His television career is legendary. In the 1950s and 1960s, he hosted a news and public affairs program in which he presented the news, weather, advertisements and interviews entirely without a script. (Courtesy of Roger Jamieson)

of the stations were owned by the Canadian Broadcasting Corporation, which operated studios in St. John's, Gander, Grand Falls, Corner Brook, Labrador City and Goose Bay.

TELEVISION

Television came to the province in September 1955, when CJON TV started transmission in St. John's on Channel 6. The station was established by the Newfoundland Broadcasting Company with studios located on Prince of Wales Street. Oscar Hierlihy, who had resigned from Avalon Telephone after a twenty-one-year career, was chief engineer in charge of technical operations. In 1991, Hierlihy was made a member of the Order of Canada for his contribution to communications in the country.

In 1956, TV station CFLA in Goose Bay started up as a joint effort of the Canadian and United States government to serve the Goose Bay Air Base. The United States Air Force provided equipment and maintenance while the CBC provided station management. The station used kinescopes, a predecessor to the VCR machine, to provide prerecorded programs from the CBC and the major US networks. The kinescopes were made in Canada and the United States and were received for rebroadcast over the station days and sometimes weeks after the original broadcast. CBC took over this station in 1964.

In 1959, the Canadian Broadcasting Corporation established its first TV station in Newfoundland when it set up CBTY in Corner Brook. In February of the same year, Newfoundland Broadcasting Company established TV station CJCN in Grand Falls, which was linked to CJON TV in St. John's by microwave facilities, creating one of the first private TV networks in Canada. In its early years, CJON TV broadcast CBC's national programming as well as local weather, news, sports and entertainment. This arrangement ceased in October 1964 when the Canadian Broadcasting Corporation opened its own TV station in St. John's and began broadcasting its own material. CJON TV subsequently became affiliated with the national CTV television network and began broadcasting some CTV national programming. CJON TV also introduced colour programming in October 1966, with CBC TV following in January the following year. CBC TV in St. John's relocated from Duckworth Street to a new facility on the Confederation Parkway in 1966.

In 1968, the CBC took over the TV station in Churchill Falls, and in 1973 acquired the Iron Ore Company of Canada's TV station in Labrador City. In 1975 the first French language TV operation in the province was inaugurated by CBC at Port au Port. Since beginning operations, both CBC and CJON TV have installed numerous TV rebroadcast stations, providing coverage to all parts of Newfoundland and Labrador. CJON in fact was the first North American television station to use a satellite television transmitter when in 1957 it put CJOX Channel 10 in Argentia on the air.

CABLE AND SATELLITE TELEVISION

Cable TV arrived in Newfoundland in August 1977 when Bay St. George Cablevision began broadcasting taped programs. The first direct programming cable system was set up by Shellbird Cablevision Company in Deer Lake in September 1977. TV programming from Canada and the United States was very costly for cable companies to bring to Newfoundland because of the high expense of renting microwave channels to carry TV signals. To share the expense, the cable TV companies across the island set up a consortium to bring mainland television channels to the province. Nowadays, mainland TV channels are received by satellite at a much lower cost, making cable TV economically viable to all consumers.

Avalon Cablevision was established in St. John's in 1977 and renamed Cable Atlantic in 1989. Cable Atlantic became the largest cable TV operation in the province, providing service to St. John's, Gander, Grand Falls, Corner Brook, Witless Bay, Port aux Basques, Isle aux Morts, Pasadena, Wesleyville, Botwood, Musgrave Harbour and Carmanville. In 2000, Cable Atlantic's cable operation was purchased by Rogers Communications Limited, and its telecommunications division by Group Telecom.

The N1 Cable System was started up in 1986, providing cable TV service to many smaller communities around the province including Stephenville, Marystown, Grand Bank, Gambo, St. Anthony, La Scie, Hare Bay, St. George's and Hawke's Bay. This company eventually became Regional Cablesystems Limited, serving mainly smaller rural communities in the province. The company operates four divisions across Canada, and in 2002 had approximately 250,000 customers.

Today, most communities in the province have a cable TV system, typically providing a choice of dozens of channels showcasing news, movies, sports and other entertainment. Satellite television is also available from two national companies, namely Bell Canada's ExpressVu and Starchoice. These satellite services provide hundreds of television channels, covering a wide range of interests.

Petty Harbour Hydro Plant
The Petty Harbour hydro plant was established by the Reid Company in 1898 with the purpose of providing power for the St. John's Street Car Company as well as for residential and commercial use in the city. This project was one of the early "engineered" projects in Newfoundland and employed several engineers. Reid engineer George Massey surveyed the route for the transmission line to St. John's. William Mackay, son of Alexander, was responsible for installing the generating equipment; and the penstock, using 3/8 inch steel, was built by the Angel Engineering Company of St. John's. The original installation had an open-top flume but was replaced by the penstock several years later. The Petty Harbour station originally produced 600 kilowatts of power and went on-stream in 1900. In 2002, the station was producing 5.25 megawatts. (Courtesy of NP)

Churchill Falls
Before being diverted as part of the mammoth hydro project, the falls were 245 feet high, or 69 feet taller than Niagara Falls in Ontario. (Courtesy of NLH)

Chapter Three

Electricity - From Petty Harbour to the Mighty Churchill

BEFORE ELECTRICITY

The modern convenience of electric power was originally stimulated by the public's need for better forms of lighting. Beginning with the early sources of lighting from fire, mankind has always sought cheaper and more efficient forms of illumination. It was no different in Newfoundland in the latter part of the nineteenth century when residents first heard of Thomas Alva Edison's invention of the incandescent lightbulb. Electric lighting did not begin with Edison's invention in 1879, as the electric arc lamp had been used from the mid 1800s. Prior to electric lighting, the staple form of artificial light was produced by kerosene, candles and oil; however, in larger cities gas lighting became available. In 1844, coal gas was introduced in St. John's for lighting purposes, just four years after it was introduced in Montreal.

In that year, the St. John's Gas Light Company built a plant at Riverhead, at the western end of St. John's, to manufacture gas by burning coal. This process also produced such by-products as coke, which was used for fuel, and coal tar, which was used as a water-proofing compound. The company installed underground gas lines along Duckworth and Water Streets, and provided lighting (and in several

---

Electric Lamps in Operation in St. John's as of Nov. 1, 1886
(Government Engineer's Office - November 6, 1886)

<u>Water Street</u>
    Patrick Street
    Jobs Bridge
    Tessier's
    New Post Office
    Beck's Cove
    Opp. P. Hutchings
    Foot of Prescott
    Foot of Cochrane
<u>Duckworth Street</u>
    McBride Hill
    Market House
    King's Beach
    Ordnance Yard
<u>Central Fire Station</u>

<u>Gower Street</u>
    West End Fire Station
    Waldegrave Street
    Callahan and Glass
    C. of E. Cathedral
    Foot of British Square
    Fort William
<u>Military Road</u>
    Garrison Hill
    Rawlins Cross
    Colonial Building
    Cochrane Street
<u>Barnes Lane</u>
<u>Foot of Signal Hill</u>

---

cases heating) for almost three hundred shops, forty wharves and twenty-five houses. The St. John's Gas Light Company continued to extend its facilities to other parts of the city. However, the company gradually abandoned the gas lighting business after electricity was introduced to the city in 1885, but continued to provide gas for heating purposes up until the late 1940s.

    During the late 1800s, lighting was typically provided in communities outside St. John's by kerosene and other oil products. In Harbour Grace, however, the Harbour Grace Gas Light Company began to provide gas street lighting in August 1852. The plant burned down in 1860 but was rebuilt and continued to provide lighting in Harbour Grace until the 1890s.

The St. John's Gas Light Company at the western end of St. John's harbour, circa 1910

EARLY ELECTRICAL UTILITIES

Electrical service for both residential and business purposes had an irregular start in Newfoundland, as was usually the case in other locations where the new technology was introduced. The early electricity producers used primarily steam or hydro to generate power which they provided to customers within a small local area. Some businesses generated electricity to operate their plants and often provided excess power to their employees and nearby residents. Over time, responsibility for power generation and distribution was passed over to electrical utilities which specialized in the business.

In Canada, the Montreal Harbour Commission was the first to use electricity for lighting when it installed arc lamps along the waterfront in 1877. Newfoundland would not see electric lighting until eight years later.

ST. JOHN'S ELECTRIC LIGHT COMPANY

The St. John's Electric Light Company (St. John's Electric) was set up in 1885 by Alexander M. Mackay to provide electric service to the residents of St. John's. Mackay was a prominent player in developing telegraph facilities in the colony, and at the time was head of the Anglo-American Telegraph Company in Newfoundland. St. John's Electric built a direct current steam-driven generator at Flavin's Lane near Rawlin's Cross in the central part of the city. The station began service on October 16, 1885, providing electrical power to eleven downtown businesses and lighting a number of carbon arc lamps. However, the inauguration of the new service was not without mishap. The electrical wires and telegraph lines shared the same poles, and the two came into contact with one another, causing a short circuit which resulted in the first power outage in Newfoundland. Repairmen fixed the problem, and a couple of days later, electric power was restored.

By 1886, electric lighting had been installed in forty-six St. John's businesses. In the same year, St. John's Electric placed twenty-five street lights for the City Council. The following year, the company installed electric lights in Government House, the Colonial Building and several other public buildings. Around the same time, it also began a major thrust to install electric lighting in private residences. This was temporarily halted in 1892, when a devastating fire wiped out most of St.

John's, including the Flavin Street power station. St. John's Electric responded by quickly erecting a new brick building at the corner of Bond and Flavin Streets and installed a steam-powered plant.

REID NEWFOUNDLAND COMPANY

The St. John's Street Railway Company (St. John's Railway) was formed in 1896 and given a fifty-year monopoly to provide electric streetcar service to St. John's. The company was set up by Robert Gillespie Reid, who two years later would organize a company to provide railway service across the island (see Chapter Four). In 1901, he formed the Reid Newfoundland Company (Reid Company) to manage his business interests in the colony.

St. John's Railway required electricity to power its streetcars, so in 1898 it began constructing a hydro-power facility in Petty Harbour to meet this demand. The $750,000 station, which was Newfoundland's first hydro electricity plant, was equipped with two 600 kilowatt generators producing 500 volts. The electricity produced was transformed to fifteen thousand volts and carried over an eight-mile-long transmission line to the city. A second line was provided to serve as backup in the event the other failed, making this arrangement the first use of electrical transmission line diversity in Newfoundland. The Petty Harbour station began service in 1900 and in addition to power for the city's streetcars, also provided electricity to residents and businesses. To help expand the electrical side of its business, the company purchased the St. John's Electric Light Company in 1901, including its Flavin Street power plant and distribution system.

Alexander McLennan Mackay (1834-1905)
Born in Pictou, Nova Scotia, Mackay came to Newfoundland in 1857 to repair the cross-island telegraph line. (Courtesy of PANL)

Since the Petty Harbour plant was providing more than adequate power for its operations, St. John's Railway closed down the Flavin Street station in 1902 and

> *Electrocution*
> *Newfoundland's first electrocution occurred on April 11, 1901, when Fred Wing, the engineer in charge of the Petty Harbour hydro plant, came in contact with a high voltage line and was killed instantly.*

had the building demolished. A tobacco business later occupied the location, which in 2002 is the site of a modern condominium complex. The company began to focus on providing power to the St. John's market and over the next twenty years greatly expanded service.

In 1920, the Reid Company re-organized its Newfoundland operations and set up the St. John's Light and Power Company to oversee its electrical and street car businesses. Four years later, the Reids sold the company to Montreal Engineering, which established Newfoundland Light and Power Limited (NLP) to manage their operations.

UNITED TOWNS ELECTRIC COMPANY

In 1902, John J. Murphy set up the United Towns Electric Company (United Towns) to provide electric power to the towns of Harbour Grace, Carbonear and Heart's Content. Murphy was a businessman who earned his wealth during the construction of the Newfoundland Railway, especially through his lumber mill and hotel operations at Gambo in central Newfoundland. The company's first generating station was located at Victoria and employed a Pelton 225 kilowatt hydro turbine which began producing electricity in November, 1904. Distribution lines were constructed to Carbonear and Harbour Grace, as well as across nine miles of barrens to Heart's Content. Additional generators were installed at this site in 1907 (225 kilowatt) and in 1913 (450 kilowatt). The plant is still providing power in 2002 and while the original generator was taken out of service in 1952, it remains in the powerhouse as a museum piece. Because of the historic significance of this facility, Newfoundland Light and Power turned part of the Victoria station into a museum, which attracts many visitors during the summer tourist season.

Over the years, United Towns expanded to other Avalon Peninsula communities. Yet the company was unable to supply power to St. John's because NLP held exclusive rights to this market. In 1914, United Towns acquired the Conception Bay Electric Company which was supplying electricity to the towns of Brigus, Cupids, Clarke's Beach, Port de Grave, Bay Roberts, and Spaniard's Bay. In 1923, the company built an 1.1 megawatt hydro station at Seal Cove, Conception Bay, to which it added a 2.5 megawatt unit in 1927.

In 1924, United Towns and NLP agreed that the former would not distribute power in St. John's within a four mile circumference of the General Post Office on Water Street. However, this arrangement was not very practical, and jurisdictional squabbles continued for many years.

United Towns expanded to the Burin peninsula in 1929 and provided service to the towns of Lawn and Grand Bank by a 185 kilowatt generator at Lawn, which was increased to 375 kilowatts in 1931. In 1930, power lines were built to Fortune, Lamaline, Lord's Cove, Point au Gaul, and St. Lawrence, and the following year to Burin. The company installed a 375 kilowatt generator at Little St. Lawrence in 1939, a 745 kilowatt hydro turbine at West Brook in 1942, a 250 kilowatt diesel generator at Grand Bank in 1956, and three 500 kilowatt diesels at Salt Pond in 1964.

The Wabana Light and Power Company incorporated in 1928 to serve the residents and mining operations on Bell Island. In 1931, United Towns purchased this company and the following year also acquired the assets of the Public Service Electric Company Limited which had been set up in 1917 to provide electricity to the northern towns of Conception and Trinity Bays. The purchase also included the hydro generating plant at Heart's Content which was originally built in 1917. In 1937, United Towns installed a 185 kilowatt diesel unit at Argentia, and in 1943, two 100 kilowatt diesel units at Freshwater.

In 1944, the company expanded to the west coast of Newfoundland when it set up a subsidiary named the West Coast Power Company (West Coast Power). In December of that year, West Coast Power obtained a fifty-year monopoly for the provision of electricity to Stephenville, Stephenville Crossing, St. George's, Port aux Basques, Channel, and several other communities on the west coast. Electricity to Corner Brook was still being provided by the local paper mill from its hydro facility at Deer Lake. In 1945, West Coast Power completed a twenty-nine hundred kilowatt hydro development at Lookout Brook, and by the end of December was supplying electricity to St. George's. In 1945 likewise, the company built a diesel plant at Port aux Basques to serve both that town and Channel. In 1948, Aguathuna and Port au Port were also added to the system. In the early 1950s, the company also provided power to the Harmon Air Force Base at Stephenville, which had outgrown its own generating capability. In 1954, an auxiliary diesel plant was installed at Dribble Brook to augment the electrical requirements at St. George's and Stephenville.

United Towns completed a 4.2 megawatt development at New Chelsea in 1957, and in 1960 replaced the Heart's Content power station. By 1965, the company was serving more than thirty-seven thousand customers. The following year, United Towns and several other smaller utilities merged with Newfoundland Light and Power.

UNION ELECTRIC LIGHT AND POWER COMPANY

The Union Electric Light and Power Company (Union Electric) was established in 1916 to provide power primarily for the operations of the Fishermen's Protective Union (FPU) in Port Union. Union Electric's first president was William Coaker, head of the FPU. The company's first power plant went into service at Port Union in 1918 and comprised a three hundred kilowatt hydro unit harnessing the flow of the Catalina River. Union Electric later provided electrical power to the surrounding communities, including Bonavista, Clarenville, Catalina, Trinity, Trinity East, Port Rexton, Champney's, English Harbour, and Elliston. The Port Union plant was augmented by a 300 kilowatt hydro unit in 1920 and a 110 kilowatt diesel unit in 1945. The station in 2002 still produces about five hundred kilowatts of power.

In 1953, the company purchased the Clarenville Light and Power Company (incorporated in 1933). This company operated small hydro units generating 125 kilowatts as well as a diesel-generating plant at Shoal Harbour providing electricity to Clarenville, Shoal Harbour, and Milton. In 1956, a three megawatt hydro plant was also installed at Lockston, Trinity Bay.

OTHER ELECTRICITY PROVIDERS

During the early part of the twentieth century, before public electric utilities began taking over power generation and distribution, some industrial companies built their own electric plants for their operations. These included the pulp and paper mills at Grand Falls and Corner Brook. The electric power plant operated by the mines at Wabana has already been mentioned. In the mid twentieth century, the Iron Ore Company of Canada and Wabush Mines, through their ownership in the Twin Falls Power Corporation, obtained electric power from a hydro development at Twin Falls. Companies such as these primarily produced power for their own needs, but also supplied excess electricity to nearby towns.

Anglo-Newfoundland Development Company Limited (AND)

The Anglo-Newfoundland Development Company Limited (AND) operated two hydro plants on the Exploits River. The first was constructed near Grand Falls in 1909 and initially provided approximately three thousand kilowatts. In 1911, the Albert Reed Company built a 10.5 megawatt hydro plant at Bishop's Falls to power its small pulp mill in the area which AND purchased in 1916. In 1956, AND's power distribution lines in the communities of Grand Falls, Windsor, Bishop Falls, Badger, Botwood, Millertown, and Terra Nova were taken over by NLP, followed in 1960 by the power generating system.

Newfoundland Products Corporation

In 1922, Newfoundland Products Corporation began construction of a hydro facility at Deer Lake to power its Corner Brook mill, the building of which would begin the following year. The installation included a dam across Grand Lake which raised the water level about thirty feet, and a seven mile canal to a forebay, which was 260 feet above the generating station. Seven penstocks were built to direct water to the powerhouse to turn seven 10.5 megawatt horizontal turbo generators. The facility began electrical production in 1925, and in addition to the needs of the mill, also provided electricity to the towns of Corner Brook and Deer Lake. In 1926, excess power from the station was purchased by the Bay of Islands Power Company which served customers in the communities surrounding Corner Brook. In 1929, two additional penstocks were constructed as well as two vertical generators, bringing the Deer Lake power station's output up to 114 megawatts.

In 1929, electric power from Deer Lake was extended via a new transmission line to American Smelting and Refining Company's Buchans mine site. The Bowater Pulp and Paper Mill Limited (Bowater), which had purchased the Corner Brook mill and Deer Lake power station in 1938, continued to hold on to its generating plants, but in 1951 began divesting itself of its distribution systems when it sold its power lines in

Corner Brook and Deer Lake to Newfoundland Light and Power. At about the same time, the distribution system of the Bay of Islands Power Company was also sold to NLP.

Bowater set up a subsidiary named Bowater Power Company in 1955 to operate its hydro operations. The following year, the new hydro company installed a 4.5 megawatt hydro plant at Watson's Brook, near

The generators at the Deer Lake power station circa 1934 (Courtesy of CBPP)

Corner Brook, and by 1965 was providing electricity to Springdale and the Baie Verte areas. NLP had previously purchased the distribution system on the Baie Verte Peninsula in 1960. In 1967, some of the power from the Deer Lake plant was connected into Newfoundland and Labrador Hydro's power grid, but as of 2002, its seven 60 hertz and two 50 hertz generators were providing 125 megawatts of power exclusively for the Corner Brook mill.

Iron Ore Company of Canada/Wabush Mines

In 1954, the Iron Ore Company of Canada built a hydro generation plant at Menihek Lake to supply electricity for its iron ore operations at Schefferville, Quebec. A hydro plant at Twin Falls came on stream in 1962 for a short period of time to provide power for the Labrador City and Wabush mining operations and town sites, as well as for use in the construction of the Churchill Falls hydro facility.

RECENT ELECTRICAL UTILITIES

During the first half of the twentieth century, commercial electrical power in Newfoundland was provided by several small utilities scattered throughout the island. The dispersion of these utilities also meant there was no common power grid, which for all intents and purposes meant that the power companies were not electrically connected with one another. The formation of the Newfoundland Power Commission in 1964 and the consequent Bay d'Espoir development, along with the emergence of the expanded Newfoundland Light and Power in 1966, ended this era. The resulting province-wide power grid provided a more reliable service and enabled a more efficient operation of electrical facilities.

NEWFOUNDLAND LIGHT AND POWER

In 1924, after a couple of years of negotiation, Royal Securities (Royal), a Montreal investment firm, purchased the assets and rights of St. John's Electric for approximately $1.2 million, and set up Newfoundland Light and Power (NLP) to operate the utility. Royal placed the operation of NLP under the control of Montreal Engineering, one of its subsidiaries. NLP's shares meanwhile were managed by International Power Company, another Royal company, which kept control until 1949, when they were sold on the open market.

Royal began evaluating St. John's Electric in the early 1920s, during which time Montreal Engineering determined that the demand for electrical power in St. John's would soon outstrip Petty Harbour's capability. Various options were considered, but the engineers decided that Petty Harbour's capacity could be increased, avoiding the need for a new hydro development. NLP subsequently increased its output to 5.2 megawatts in 1926, and upgraded the transmission line to St. John's to thirty-three thousand volts. After a few years, electrical demand again exceeded what the plant could deliver. A new source of power was needed, and the company's engineers found this at Pierre's Brook, south of Petty Harbour. In 1930, construction began on the 3.2 megawatt hydro development, which produced electricity not only for consumers in the St. John's area, but for the mining companies on Bell Island as well.

Aaron Bailey, P. Eng. (1906-1990)
Bailey served as Newfoundland Light and Power's President between 1967 and 1978, and as Chairman of the Board of Directors between 1978 to 1985. (Courtesy of NP)

NLP also constructed several other hydro plants on the south Avalon peninsula. These included a 4.5 megawatt hydro plant at Tors Cove, which went into service in 1942, as well as a 9.35 megawatt hydro project at Mobile Big Pond, which went on line in 1950. An eighteen-mile-long transmission line to the Goulds substation was also part of the Mobile development. When the Mobile Big Pond went on line, it provided about one third of the company's annual power requirements. Other construction in the early 1950s included a 6.0 megawatt hydro unit at Cape Broyle in 1953, and a 7.65 megawatt plant at Horse Chops in the same year.

Prior to the 1930s, the mining companies on Bell Island generated their own electricity using coal-fired steam. In 1930, NLP saw the economic opportunity of serving the power-hungry mining operations and installed two 13 kilovolt cables across the tickle from Broad Cove to Bell Island for this purpose. In 1955, a new 33 kilovolt submarine line was installed to meet increased demand.

> **Presidents of Newfoundland (Light and) Power**
>
> 1966-67 Denis Stairs, P. Eng.
> 1967-78 Aaron Bailey, P. Eng.
> 1978-82 Alastair D. Cameron, M.B.E. P. Eng.
> 1982-85 David S. Templeton, P. Eng.
> 1985-89 Angus A. Bruneau, O.C. P. Eng.
> 1990-97 Aidan F. Ryan, P. Eng.
> 1997 - Philip G. Hughes, C.A.

NLP continued to expand its operations. In 1951, it purchased Bowater's distribution systems on the west coast, and in 1956, AND's distribution systems in central Newfoundland. The company continued to develop its generating capability and in 1958 constructed a six million dollar 12.75 megawatt hydro plant at Rattling Brook, followed in 1963 by a six megawatt hydro facility at Sandy Brook. In 1965, the company completed a 185 mile transmission line from Gander to St. John's, which interconnected NLP's eastern and central Newfoundland power systems for the first time.

In 1958, NLP took over the distribution systems in both Gander and Lewisporte. The Lewisporte power system had had its beginnings in 1937, when the firm of A. T. Woolfreys and Brothers provided electricity to houses in the town from excess power produced by the steam plant which they used for their logging operations.

NLP installed a 1.865 megawatt diesel generating unit on the south side of St. John's harbour in 1953. Three years later, the company opened a new facility on the south side and inaugurated a ten megawatt steam plant, followed by additional equipment over the next few years, bringing the plant's total generating capacity up to thirty megawatts in 1959. The plant continued to operate until 1999, when it was retired and dismantled.

> **Newfoundland Power's assets (as of Dec 31, 2000)**
>
> - 23 hydro plants, 5 diesel plants, 3 gas turbine plants producing a total of 148.4 megawatts
> - 141 substations
> - 10,000 km of distribution and transmission lines
> - 52,000 streetlights ($30 million)
> - 45,000 transformers ($100 million)

In 1966, NLP amalgamated with the

United Towns Electric, Union Electric Light and Power, West Coast Power, and Public Service Electric companies. The resulting organization continued under the name of Newfoundland Light & Power Co and Denis Stairs, P. Eng. became its first president. The result of this arrangement was that NLP became one of only two public utilities in the province. Newfoundland Hydro, the other, will be discussed later.

Over the next two decades NLP continued to grow and greatly expanded its distribution plant. Leading this growth were Aaron Bailey, P. Eng., who succeeded Stairs as president in 1967. He was followed in 1978 by Alastair Cameron, M.B.E., P. Eng. and in 1982 by David Templeton, P. Eng.

In 1984, the company took advantage of a Federal government program which gave a tax advantage to the construction of small hydro stations, and built its first hydro plant since the 1960s. This was an 1.1 megawatt station utilizing the Mobile River watershed on the southern shore of the Avalon peninsula. It was named the Morris power station, after John William Morris, a former General Manager of the company. The following year, Angus A. Bruneau, O.C., P. Eng. became president.

In 1987, NLP underwent a structural reorganization. Its single shareholder became Fortis Inc., a holding company which later expanded into real estate, hotel, and other electrical utility operations in Belize, Cayman Islands, Nova Scotia, Ontario, New York, and Prince Edward Island.

In 1990, Aidan Ryan, P. Eng. became Newfoundland Light and Power's president. In

Aidan F. Ryan, P. Eng.
Ryan was appointed President and Chief Executive Officer of Newfoundland Light and Power on January 1, 1990, and served in that capacity until 1997. (Courtesy of NP)

Newfoundland Power generating station at Tors Cove (Courtesy of NP)

the same year, the company underwent a corporate name change to become Newfoundland Power. (The name was legally changed to Newfoundland Power Inc. in 1998.) In 1997, Ryan retired as president and was succeeded by Philip G. Hughes, C.A.

Newfoundland Power's latest hydro installation – the first in fourteen years – went into service in 1999. This was a 6.0 megawatt development at Rose Blanche on the southwest coast, which cost approximately $13.8 million, $1 million of which was spent on environmental studies.

NEWFOUNDLAND POWER COMMISSION

In the second quarter of the twentieth century, the power utilities expanded service to many communities. Yet by the early 1950s only about fifty per cent of Newfoundland's population had access to electric power. The smaller power companies that had sprung up either fell by the wayside or were taken over by the larger utilities. Since it was very costly to provide electric power to smaller towns, the Newfoundland Government set up the Newfoundland Power Commission (Commission) in 1954 to extend electric power to rural area, and the following year appointed George Desbarats, P. Eng. as its first chairman. One of the

Commission's early initiatives was the installation of a two hundred kilowatt diesel generator at Happy Valley-Goose Bay in 1958. Desbarats stayed on until 1958 and was succeeded by John Ryan P. Eng., who in turn was followed by Frank Newbury, P. Eng. in 1961 and George Hobbs, P. Eng. in 1964.

In order to accelerate the provision of electricity to the remote areas of the province, the government created the Rural Electrification Authority in 1964 to assume responsibility for providing electric power to the province's small unserved communities. By 1972, the rural electrification program was completed, and all told, forty-six diesel plants were installed, providing electricity to all communities with more than fifteen customers.

In 1965, the Newfoundland government reorganized the Power Commission and gave it responsibility for developing major hydro developments in the province. Its first major hydro project was the Bay d'Espoir development, which began construction in 1965.

British Newfoundland Development Corporation (Brinco)

Shortly after Confederation, the provincial government recognized the mineral and hydro potential of Newfoundland and Labrador and proceeded to encourage development of these resources. In 1952, Premier Joseph R. Smallwood visited England to encourage British business interests to invest in the large areas of underdeveloped land in the province. He met with Anthony de Rothschild and Sir Winston Churchill, the prime minister at the time. Smallwood convinced the British House of Rothschild to organize a consortium of mainly British and Canadian banks and industrial firms to form the British Newfoundland Corporation. The original companies in the consortium included N. M. Rothchild & Sons, the English Electric Company Limited, Rio Tinto Company Limited, Anglo-American Corporation of South Africa, Bowater Paper Corporation Limited, and Frobisher Limited. Prior to signing the main agreement with the province of Newfoundland, seventeen other firms joined the consortium, including Bowring Brothers Limited, Suez Canal Company, the Royal

Newfoundland Power Commission Chairmen	
	Tenure
George H. Desbarats (1900 - 1999)	1956 - 1958
John C. Ryan (1925 -)	1958 - 1961
Frank Newbury (1920 -)	1961 - 1964
George Hobbs (1917 - 1988)	1964 - 1973
Wallace S. Read (1930 -)	1974

Bank of Canada, and the Bank of Montreal. The consortium eventually changed its name to the British Newfoundland Development Corporation, or Brinco for short.

As part of the agreement, Brinco was given a twenty-year concession on mineral and water rights to fifty thousand square miles of Labrador (including the Hamilton Falls watershed) and ten thousand square miles on the island part of the province. At the end of the first year, ten thousand square miles of the Labrador concession would revert back to the government, and eight thousand square miles every five years thereafter. A proportionately similar arrangement also applied to the land rights on insular Newfoundland. Brinco in return promised to spend at least $1.25 million on exploration and development every five year period, and pay the government eight per cent of its pre-tax profits.

Bay d'Espoir

One of Brinco's first areas of interest was the hydro potential of the Bay d'Espoir region on the province's south coast. In 1959 the company set up Southern Newfoundland Power and Development Limited which partnered with Bowater, Anglo-Newfoundland, and the Power Corporation of Canada (a major shareholder of United Towns) to develop the hydro potential of the area. All partners would benefit – the paper companies for electricity to power their mills, and the Power Corporation for additional power to satisfy United Towns customers in the eastern part of the province.

It was not until 1963, after the Newfoundland government obtained $20 million of financial support from the Canadian government, that Premier Smallwood gave the go-ahead to the Newfoundland and Labrador Power Commission to begin work on the project. By then Brinco was deeply involved in planning a hydro-electric development on the Churchill River in Labrador, so it decided it would not participate in the Bay d'Espoir development. It consequently sold its rights back to the Newfoundland government.

One of the perceived benefits of the Bay d'Espoir project was the attraction of new industry to the province, as it was expected that the power generated would lead to the development of major industrial complexes and spur economic growth. The Electric Reduction Company of Canada plant at Long Harbour, the Stephenville liner board mill, and the oil refinery at Come by Chance were some of the industries which started up after Bay d'Espoir power became available.

Up until the 1960s, hydro projects in Newfoundland were small in scale, with only Deer Lake greater than one hundred megawatts. Bay d'Espoir was a significant engineering development – at the time, the largest ever undertaken in the province. The first stage cost $58.4 million and produced 225 megawatts, more than twice the province's electrical generating capacity at that time. Engineers built seven dams and five canals to divert the Victoria, White Bear, Grey and Salmon rivers to the generating plant. The first stage cost $58.4 million. In 1970, at an additional cost of $63 million, the capacity was doubled to 450 megawatts by diverting the White Bear and Victoria rivers and installing new generating units. In 1977, another generating unit was added, bringing the total capacity up to six hundred megawatts. The total cost of the development was approximately $190 million.

George Hobbs (1917-1988)
Hobbs was born in Heart's Content and graduated from McGill University with an engineering degree. He worked in Ontario for five years and assumed a position with Bowater in 1946, becoming Chief Engineer in 1954. He was appointed Chairman of the Newfoundland Power Commission in 1964, and during his tenure oversaw the rural electrification program, the Bay d'Espoir development, and the Holyrood thermal plant. (Courtesy of NLH)

In conjunction with the new generating plant, a new power grid was built across the province, which was designed to carry power not only from Bay d'Espoir, but also from a potential hydro development in Labrador which was being considered at the time. A 230 kilovolt transmission line was constructed east to St. John's via Come by Chance and Sunnyside, as well as a 138 kilovolt spur between Sunnyside and Marystown. To the west, a 230 kilovolt line was also built to Grand Falls, Corner Brook and Stephenville, with 138 kilovolt spur lines to the Springdale, Codroy, and Port aux Basques areas.

The project was engineered and constructed by Shawmont Newfoundland Engineering, a joint venture of Shawinigan Engineering and Montreal Engineering. Electric power first flowed from the development in 1967. Although Bay d'Espoir was a large engineering undertaking, it would be dwarfed several years later by a major hydro development in Labrador.

Frequency Conversion

Prior to the Bay d'Espoir development, electricity on the Avalon peninsula and eastern portion of the island was generated at a frequency of sixty hertz, while power generated elsewhere was primarily fifty hertz. Fifty hertz power was produced because the equipment at Grand Falls and Corner Brook mills operated at that frequency, reflecting the English standard adopted by the companies which originally built the plants. Initially, it was planned to install four 52.2 megawatt hydro generators at

Bay d'Espoir hydro-electric plant
Construction on the 450 megawatt hydroelectric development began in 1965, and the first power was produced in 1967. (Courtesy of NLH)

Hydro transmission lines over Grand Lake in western Newfoundland
The first phase of this project was completed in 1967. The span was almost eighty-seven hundred feet long, one of the longest overwater lines in the world at the time. (Courtesy of NLH)

Bay d'Espoir – two units operating at fifty hertz for the two paper mills, and two at sixty hertz for consumers in eastern Newfoundland.

In 1964, the government set up a committee to look into the pros and cons of whether Bay d'Espoir should produce power at one or both frequencies. After some consideration, it decided to adopt the North American standard and produce only sixty hertz power. Plans for the Bay d'Espoir project were changed to three 75 megawatt sixty hertz units, rather than the four 52.2 megawatt generators (two at fifty hertz and two at sixty hertz) which had been previously contemplated.

A province-wide program was implemented in 1964 to standardize on sixty hertz. Frequency converters were installed at Grand Falls and Corner Brook to accommodate the paper companies and towns, as well as the Whalesback Mine at Springdale. The Atlantic Development Board provided $4 million for the conversion project.

> **Bay d'Espoir Construction Facts**
> - *The surge tanks on the penstocks, at 371 feet above their base, were the highest in the world.*
> - *The 230 kilovolt transmission grid employed the highest voltage east of Quebec.*
> - *The eighty-three hundred foot transmission line span across Grand Lake was the longest at its voltage in Canada.*
> - *Meelpaeg Lake, the man-made reservoir constructed for the project, is now the third largest on the island.*

Holyrood

Even with the large amount of power generated at Bay d'Espoir, the promotion of electric appliances and heating was causing an unexpected increase in the province's forecasted electrical load. To meet this increased demand, the Commission's engineers decided to install an oil-fired steam plant at Holyrood. The management and engineering of this project was done by Shawmont Resources, a joint venture of consultants. Two 150 megawatt turbine units were put in place, one in 1970, and the other in 1971, at a total cost of $48.2 million. A third turbine was installed in 1979, bringing the generating capacity of Holyrood up to 450 megawatts, almost half of the electrical power produced on the island by Newfoundland and Labrador Hydro (the Commission's successor). The Holyrood complex also required the construction of a marine terminal to accept delivery of the Bunker C oil which was used to fire the plant's boilers, as well as new transmission lines to St. John's. By 2002, after undergoing various improvements, the station was producing approximately 500 megawatts.

Churchill Falls

Grand Falls on the Hamilton River in Labrador was first reported by a European Canadian in 1839. John McLean, of the Hudson's Bay Company, came upon the waterfall when he was searching for a route to Fort Chimo, an outpost on Ungava Bay. The region remained in relative obscurity until iron ore was found west of the area about half a century later, the falls remaining only of geographic interest until its hydro-electric potential became realized in the mid 1900s. In 1965, Grand Falls and the Hamilton River were renamed Churchill Falls and Churchill River respectively, in honour of Sir Winston Churchill who had died earlier that year.

The falls are about 190 miles west of the Churchill River drainage basin at Lake Melville, and plummet more than 245 feet. The system feeding the waterfall drains about twenty-two thousand square miles of western Labrador. The Upper Churchill, in less than twenty miles, drops about one thousand feet with a water flow volume of thirty to forty thousand cubic feet per second, creating a huge hydro electric potential.

Brinco had a strong interest in developing the area and contracted with Shawinigan Engineering Company Limited and Montreal Engineering Company Limited to survey the hydro potential of the Hamilton River. They concluded that the development of the river's hydro resource would be one of the largest such projects in the world. In 1958, Brinco decided to proceed and set up the Hamilton Falls Power Corporation to start work on the project.

The Holyrood thermal plant in 2002

Electric Reduction Company of Canada Industries Limited

One of the companies that took advantage of the Newfoundland government's incentive of low commercial power rates from the Bay d'Espoir development was Electric Reduction Company of Canada Industries Limited (ERCO). ERCO's business was the production of phosphorous, a process which required massive amounts of electricity. With government incentives, the proximity to the power grid, and a convenient harbour, ERCO decided to set up a plant at Long Harbour, Placentia Bay. The facility went into production on December 1968, using power from Bay d' Espoir costing 2.5 mills per kilowatt hour. Phosphate rock, the principal material for the plant, came mainly from Florida, but other supplies came from Canada, as well as silica from Dunville, Placenta Bay.

*The liquid phosphorous was shipped in specially designed ships (**Albright Pioneer** and **Albright Explorer**) to various parts of the world. By 1980, the plant was annually producing more than forty thousand tons of phosphorous, and ERCO was employing about 450 people. Because of pollution and environmental concerns however, the plant was forced to close in 1986.*

In 1961, the Newfoundland government granted the Hamilton Falls Power Corporation – later known as the Churchill Falls Labrador Corporation (CFLCo) – a ninety-nine year lease for the development of the hydro capability of the upper Churchill River. The impediments to developing the upper Churchill River's hydro electric potential were numerous: among them were the engineering problems of building the power plant, water feeder structures, and transmission lines to carry the electrical power over the long distance to market; the remoteness of the area; the high cost of construction; finding an end customer to purchase the power; and securing satisfactory financing arrangements. It was not until after the Iron Ore Company of Canada began development of the iron ore deposits in western Labrador that serious consideration was given to harnessing the Upper Churchill's hydro capability. The Quebec North Shore and Labrador Railway, built in 1952 to transport iron ore from Schefferville to Sept-Iles, on the Gulf of St. Lawrence, became a critical component of the Churchill Falls project. In 1952, Brinco constructed an access road from the Churchill Falls area to the train line at Esker, about 125 kilometres to the west. This connection allowed heavy equipment and materials to be shipped to the hydro site. An initial part of the Churchill Falls development was the construction of a hydro generating plant at Twin Falls, located on the Unknown River, a tributary of the Churchill, west of Churchill Falls. The facility went into service in 1962 and produced 225 megawatts of electricity. Its power was used not only for the iron ore developments at Wabush and Labrador City, but also for Churchill Falls construction a few years later. After Churchill Falls came on stream, engineers found it more expedient and economic to divert the water feeding the Twin Falls generating plant to the main Churchill Falls plant, so Twin Falls was subsequently decommissioned. Nevertheless, it is still being maintained for possible future use.

> ***The Churchill Falls Development:***
> - *At the time of construction, the largest civil engineering project undertaken in North America*
> - *27,000 square mile catchment area*
> - *49,000 cubic feet per second water flow*
> - *296.3 metre long, 24.7 metre wide, 46.9 metre high underground power house*
> - *340,000 cubic yards of concrete used*
> - *$800 million construction cost*
> - *5225 megawatts of power generated (uprated in 1985 to 5428.5 megawatts)*

Construction of the Churchill Falls site began in 1966, and by year's end, a main camp and two construction camps had been built. One of the first orders of business was the construction of an airstrip to facilitate the transportation of personnel and light equipment. The heavy materials were shipped by rail to Esker and then by road to the construction site.

Above the falls, engineering plans called for the rivers to be dammed, canals built, and more than forty miles of dykes constructed, to create Smallwood reservoir, a man-made lake of twenty-seven hundred square miles, about half the size of Lake Ontario. Engineers decided that the control structure to regulate the flow of water would be located at Lobstick, about forty-eight miles upstream from the falls and sixty-four miles upstream from the powerhouse. The three control gates at Lobstick were operated remotely from the main control station at the powerhouse.

Donald J. McParland (1929- 1969)
McParland graduated with a degree in mechanical engineering from the University of Toronto in 1952. After graduation, he joined Rio Tinto, which in 1963 entered into a management agreement with Brinco. In 1966, he became President and Chief Executive Officer of Churchill Falls (Labrador) Corporation. He was tragically killed in a plane crash near Wabush in 1969. (Courtesy of NLH)

The underground complex was cut out of solid rock about one thousand feet below the top of the falls. It consisted of eleven penstocks, powerhouse, transformer gallery, bus tunnel, surge chamber, and trail race tunnels. The water was delivered via the eleven 427 metre long concrete penstocks to the powerhouse, feeding eleven vertical-shaft Francis turbine waterwheels. Each turbine was eighty-five tons and nineteen feet in diameter, creating a power of approximately 485 megawatts – collectively, a power-generating capability of 5225 megawatts (increased to 5429 megawatts in 1988).

| Challenge and Change | - 74 - | Electricity |

Churchill Falls Powerhouse
Carved out of rock one thousand feet below ground, this powerhouse is an engineering marvel. At 296.3 metres long by 24.7 metres wide, it is the largest underground powerhouse in the world. (Courtesy of NLH)

     The voltage was increased in the transformer gallery and taken to the surface via the bus tunnel. The electric power was then delivered to the Quebec Hydro grid 126 miles away. The transmission line was constructed mainly during the winter season when the muskeg was frozen. The control and administration buildings were on the surface, and along with the reservoir and transmission lines, were the only sections of the Churchill Falls development visible above ground.

The Churchill Falls project was overseen by Donald McParland (1929-1969), P. Eng., who had obtained a degree in Mechanical Engineering from the University of Toronto in 1952. McParland became VP (Technical) of Brinco in 1963 and after becoming President of CFLCo

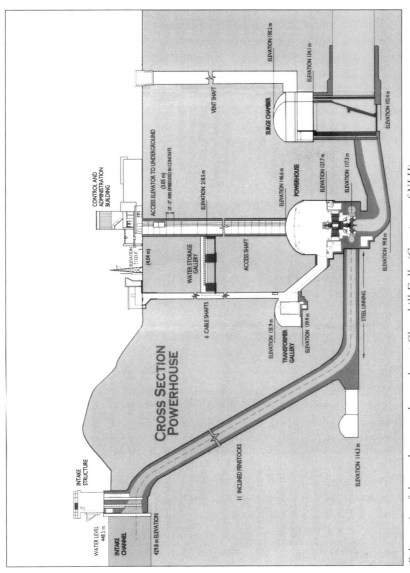

Schematic of the underground complex at Churchill Falls (Courtesy of NLH)

in 1966 succeeded to the Brinco presidency in 1969. Tragically, he was killed in an aircraft accident, when his business jet crashed short of the runway near Wabush on November 11 of the same year.

The day-to-day overall general project management was provided by Acres Canadian Bechtel (ACB). The overall project required eighty major contracts, involving about fifty different companies.

The Churchill Falls town site, located at the upper end of the Falls, was closely controlled by CFLCo and fully designed to take the needs of a remote community into account. A road was built between the town and Happy Valley-Goose Bay, and an airline service to the site was inaugurated in 1970. By 1972, the town had a recreational centre, hospital, bank, barber and beauty shops, library, hotel, bowling alley and movie theatre. At the peak of construction, some sixty-three hundred employees worked on the $950 million Churchill Falls project. Power began flowing to Hydro-Quebec on December 6, 1971.

The Churchill Falls development has been at the centre of a political controversy in Newfoundland and Labrador for decades. In order for construction of the development to commence, there had to be a guaranteed customer for the power. Hydro-Quebec was this customer, which through a number of complicated legal agreements agreed to purchase the power at a fixed rate, with no built-in clause for inflation. The 1960 rate is extremely low relative to twenty-first century power rates, and apart from the construction jobs, the province has benefited little from the agreement. Nonetheless, in spite of legal challenges, the original agreement will stay in effect until 2041.

NEWFOUNDLAND AND LABRADOR HYDRO-ELECTRIC CORPORATION

In 1974, the Power Commission reorganized and became Newfoundland and Labrador Power Corporation (NLPC). Although still a crown corporation, NLPC retained its own Chairman, Board of Directors, Chief Executive Officer and executive, and was run as an independent operation. NLPC also took over the Churchill Falls Corporation at the time.

NLPC's first chairman was Dennis J. Groom, and its first president was Wallace S. Read, P. Eng. The following year, the organization again restructured and became Newfoundland and Labrador Hydro-elec-

Chairs of Newfoundland and Labrador Hydro	
Douglas Fullerton O. C., (1917-1996)	1975
John C. Crosbie (1931 -)	1976
Denis Groom (1932 -)	1976 - 1978
Victor L. Young (1945 -)	1978 - 1984
Cyril Abery (1936 -)	1985 - 1991
James R. Chalker (1931 -)	1992 - 1996
Norman J. Whalen (1945 -)	1996 - 1999
Dean T. MacDonald (1959 -)	1999 - present

tric Corporation (Hydro). Douglas Fullerton became Chairman of the Board of Directors, and Groom became President and Chief Executive Officer, while Read became senior Vice-President. The government received dividends from Hydro's earnings and was its sole shareholder. Hydro was, and still is, regulated by the Public Utilities Board of Newfoundland with respect to its spending and as well as the rates it charges customers.

In 1975, Hydro installed a fifty-four megawatt gas turbine at St. John's and the following year a similar unit at Stephenville. In 1977, it began construction of a hydro project at Hinds Lake, in central Newfoundland between Deer Lake and Buchans. This seventy-five megawatt plant was completed in 1980. A 138 kilovolt transmission line was constructed to Howley, where it connected into the province's electrical grid. The main contractor for this $85 million development was McNamara Corporation of Newfoundland and the project manager was ShawMont-Lavalin.

An eighty-four megawatt hydro development on the Upper Salmon River, approximately thirty miles west of Bay d'Espoir, was placed in service in 1983. The project took advantage of the undeveloped watershed upstream from the Bay d'Espoir hydro plant.

In 1980, the province implemented the Environmental Assessment Act. Newfoundland and Labrador Hydro, at the time, was planning a development at Cat Arm, on the Great Northern Peninsula. This project was one of the first to be approved under the new Act's provisions. Construction on the $311 million project began in 1981. The installation of two 63.5 megawatt Pelton turbines was completed in 1984 and the plant became operational in 1985. Power was carried via a 230 kilovolt line to the main power grid and was remotely controlled from Hydro's control centre at Bay d'Espoir.

Hydro acquired Wabush's electrical distribution system in 1985, and seven years later, Labrador City's. In 1989, the company constructed a $28 million five megawatt wood chip steam plant at Roddickton, on the

Cat Arm Hydro Development
The $85 million hydro development on the Great Northern Peninsula was constructed for Newfoundland and Labrador Hydro by McNamara Construction. The 120 megawatt project, which opened in 1985, used 8000 cubic metres of concrete and required the construction of thirty kilometres of access road. (Courtesy of NLH)

Great Northern Peninsula, and a sixty-six kilovolt line to St. Anthony. This plant was in addition to a 425 kilowatt hydro plant installed there in 1981. Roddickton was connected into the provincial power grid in 1996, at which time the wood chip plant was taken out of service. Hydro also built a $24 million hydro plant at Paradise River on the Burin peninsula in 1989 which produced eight megawatts of power. In 1992, Hydro installed a 27 megawatt gas turbine at Happy Valley-Goose Bay as a backup for its existing power from a 138 kilovolt line from Churchill Falls.

In June 2001, work began on the Granite Canal hydro facility, which is located about eighty kilometres northwest of the Bay d'Espoir power plant. The engineering design and project management for the major power structures were carried out by AGRA-BAE Newplan Joint Venture. The $135 million facility is expected to begin service before August 2003.

Paradise River, Placentia Bay hydro development
Built in 1989, this plant generates eight megawatts. (Courtesy of NLH)

Star Lake Hydro Development
This 18.4 megawatt development was the first to be constructed by private interests with the express purpose of providing electricity to the provincial power grid. The $34 million plant was constructed by McNamara Construction for the Star Lake Hydro Partnership/CHI Hydroelectric Co./Abitibi Price Inc. Power generation started in October 1998. (Courtesy of McNamara Construction)

Installed Generating Capacity, Electric Utilities and Industrial Establishments in Newfoundland and Labrador (kilowatts)

YEAR	HYDRO	THERMAL	TOTAL
1956	206120	28549	234669
1960	257430	56264	313694
1965	461445	69185	530630
1970	874116	273902	1248018
1975	6205766	462164	6748227
1980	6444256	750419	7194675
1985	6559655	756660	7316315
1990	6649786	811838	7461624
1995	6647676	767754	7415430
2000	6691398	725457	7416155

Source: Statistics Canada, 57-202, 57-206

St. John's Streetcars
The original St. John's streetcars went into service in May 1900. The streetcars were electrically-powered and had a maximum speed of eight mph. They held forty to fifty passengers, and were operated by a conductor and brakeman. In the 1920s, new streetcars were introduced that required only one operator and that could attain a speed of twenty mph. The main depot was at Water Street west, near the junction of Old Topsail Road. The track ran east on Water Street to Holloway Street, then up to Duckworth Street, and east to the Newfoundland Hotel, then west on Military and Queen's Road, down Adelaide Street to the terminal. The streetcars were abandoned in favour of a bus system in 1948. (Courtesy of the A. C. Hunter Library, Newfoundland Collection)

Chapter Four

Transportation - Land, Sea and Air

Newfoundland's transportation history, like that of most islands, is tied to the sea. Practically all of the province's towns are located on the coast, and transportation between them was traditionally by boat. Where communities were close together, they were connected by footpath. Later, roads and bridges evolved, first for horse and carriage, and in the twentieth century, for motor vehicles. In the mid 1800s the arrival of the steam engine provided mechanical propulsion to shipping, and also introduced a new form of transportation – the railroad – which first appeared in Newfoundland in the mid 1880s.

In the 1900s, gasoline, diesel and jet engines ushered in new modes of transport, namely automobiles, trucks and aircraft. The evolution of motor transport, for both the movement of people as well as the movement of freight, ore, oil, fish, wood and other resources – along with the highways, bridges, ports and airports to accommodate these vehicles – constitutes an important aspect of Newfoundland's engineering heritage and is the focus of this chapter.

HIGHWAYS

Roadways, along with houses, basic tools and utensils, are among the earliest of engineering construction projects. Among the first streets in Newfoundland were those at Lord Calvert's colony at Ferryland, which date from the 1620s. This was one of the earliest communities in Newfoundland, the excavation of which has revealed cobblestone streets. In St. John's, Water Street is considered by many to be the oldest street in North America. That may be the case, but it certainly began as a footpath. As permanent houses and settlements evolved, streets were built to connect houses, and roads to connect settlements. The primary focus of the following discussion will be on roads and highways between communities.

> **Governor Cochrane's Speech to the Commercial Society of St. John's - October 31, 1831 (excerpt)**
>
> "Gentlemen, on my arrival among you, there was nothing more than the trace of a road one hundred yards from the town - where the King's Bridge now stands, but a beam lay across by which passengers tremblingly passed to the other side; on visiting Portugal Cove I was under the necessity of leaving my English horses at Windsor Lake and of proceeding the remainder of the way upon a country horse; the road to that place is now in quality and beauty, almost equal to any in England, with two stages upon it every day."
> - ***Public Ledger***, October 31, 1831

In Newfoundland and Labrador's early days, roads were few and far between. They began as footpaths, some of which could be traversed by horse and sled or carriage. Though the earliest mode of travel between communities in Newfoundland was via rowboat, skiff or sailing vessel, road construction appeared early in the history of the island. In the eighteenth and nineteenth century, the British military constructed roads between their forts in St. John's, among them Military Road between Fort Townshend and Fort William.

However, apart from military construction, almost none of Newfoundland's early roads between towns were planned. One of the first references to a road appeared in government crown lease records which mention a road leading to Torbay in 1803. It was not until Governor Cochrane arrived in the colony in 1825 that some semblance of planning began and the government began to allocate funds for road construction. The first road designed to accommodate horse and carriage traffic in Newfoundland was built between St. John's and Portugal Cove in 1825. Other roads were soon constructed to link St. John's with towns around Conception Bay and with other communities on the Atlantic coast

On the Humber River near Corner Brook, from the top: 1896, 1938, and 1996 (Courtesy of the Department of Works, Services and Transportation)

south of the city. These were far from the engineered paved projects that we would expect today, but rather rough gravel paths cleared from the forest, for the most part, only passable during dry conditions or during the winter when they were frozen. Beginning in the 1820s, the government hired surveyors to begin examining routes for roadways, which were at the time designed for only a width of nine feet, hardly a road by today's standard. Regardless, during the 1800s, the government constructed over three thousand miles of road, of which about one thousand were built for the postal service.

The twentieth century saw the introduction of motor vehicles, the first of which arrived in Newfoundland in 1903. With the number of motor vehicles rising, modern highways became a necessity, so in 1925 the government set up the Newfoundland Highroads Commission to oversee road construction in the colony. By the early 1930s, many communities on the Avalon peninsula were connected by roads. Many other communities in the colony also had road systems, but for the most part, these towns were not connected with one another. The towns with the best roads were the "company towns" built by mining and paper companies, most of which were planned communities, possessing extensive road structures.

Between 1900 and 1934, the Newfoundland government spent $15 million on road construction and repair, but there was little of either during the economic depression of the early 1930s. In 1934, the Commission of Government commenced some road construction. A thirty-mile road was built between Corner Brook and Deer Lake, as was an eighteen-mile stretch between Grand Falls and Badger. These were constructed under engineering supervision and were the best stretches of highway in Newfoundland at the time. During World War Two, the RCAF built a gravel highway between Gander airport and the coastal community of Lewisporte. Newfoundland's first paved highway was a twelve mile section between St. John's and Topsail, which was completed in 1945.

In 1947, the government and the paper companies at Corner Brook and Grand Falls agreed to connect their respective company road systems via a sixty-eight mile long highway, which completed the gap between the two towns. The government contributed 50% of the cost of construction, while the companies split the other half. The project was supervised by government engineers and employed equipment owned by the paper companies. The road was completed in 1949.

On June 24, 1947, a highway was opened between Clarenville and Bonavista. This highway, known as the Cabot Trail, was in commemoration of the 450th anniversary of John Cabot's discovery of Newfoundland. Around the same time, other highways to the Burin and Bonavista peninsulas were also completed. During the period of Commission of Government (1934-1949), only about $25 million was spent on roads. After Confederation, road construction was accelerated. By 1957, more than $35 million had been spent to modernize the province's highway system, especially to the outports, where construction was difficult and costly because of the many bays and inlets. A start was also made on the Trans-Canada Highway. In the 1950s, a highway was also constructed to provide road connection to the Great Northern Peninsula, linking St. Anthony with Deer Lake. In 1958, the government implemented a $58 million, four-year program for highway improvements. By the end of 1962, an additional 627 kilometres of road had been paved, and 937 kilometres of new gravel roads built. The major construction during this period was the 390 kilometre coastal highway between Bonne Bay and St. Anthony, which Premier Joseph Smallwood officially opened on November 1 1961.

Great strides have been made in highway improvements since Confederation. In 1950, 195 kilometres of the province's highways were paved and 2897 kilometres were gravel. The 1970s saw a renewed paving program when more than three thousand kilometres of additional pavement were laid.

In the 1970s, new road construction connected Baie d'Espoir on the south coast with central Newfoundland. The Burgeo highway was also completed, linking Burgeo on the southwest coast to the Trans-Canada Highway near St. George's. In 1979, the $52 million harbour arterial highway linking the Trans-Canada Highway to St. John's harbour was also completed, followed two years later by the St. John's cross-town (Columbus Drive) arterial highway.

As of 2000, 1947 kilometres of the province's highways were gravel, and 6990 kilometres were paved.

Alexis River bridge and causeway, Labrador
The bridge on this structure used the cantilever method of construction and is the largest of its type in the world. The bridge and substructure design was carried out by the Department of Works, Services and Transportation. (Courtesy of the Department of Works, Services, and Transportation)

THE TRANS-CANADA HIGHWAY

After the province joined Canada in 1949, plans for extending the Trans-Canada Highway (TCH) system to Newfoundland began. At the time there was no highway connecting Port aux Basques on the province's west coast and St. John's on the east. The main gaps were eighty-nine miles between Port aux Basques and St. George's, thirty-four miles between the Gander River and Bishops Falls, and eighty-five miles between Gander and Clarenville.

The Newfoundland portion of the Trans-Canada Highway was originally scheduled for completion by the end of 1956. However, because of changes in financial arrangements between the Federal and Provincial governments, it was not finished until 1965. In January 1964, both governments finally agreed to a financing plan which would accelerate the completion of the highway. Newfoundland's Premier Smallwood heaped praise on then Prime Minister Lester Pearson by posting billboards along the highway exclaiming "We'll finish the drive in '65 thanks to Mr. Pearson." The commitment was met and the last section of highway was completed on November 27, 1965, twenty miles east of Deer Lake. In the previous three years, 219 miles of highway had been constructed, quite an engineering and construction undertaking considering the challenging terrain. Up to that point in time, the Trans-Canada Highway was one of Newfoundland's major engineering achievements, costing about $120 million.

The Trans-Canada Highway was officially opened by Premier Smallwood and Prime Minister Pearson on July 12, 1966, at a ceremony in central Newfoundland.

Trans-Canada Highway Contractors

Contractor	Constructed (miles)	Paved (miles)
Chisholm Construction	27.7	15.0
Concrete Products	108.4	-
Curran & Briggs	22.1	42.9
J. Goodyear & Sons	-	59.2
Lundrigan's	131.3	66.9
M. A. Rose & Son	5.5	-
McNamara Construction	91.7	168.9
Modern Construction	32.6	-
Trynor Construction	14.1	39.7
Western Construction	111.4	145.5

From *Highway to Progress* (1966)

The Dildo Run causeway (Courtesy of the Department of Works, Services and Transportation)

THE TRANS-LABRADOR HIGHWAY

In 1997, the provincial government committed $190 million for the construction and improvement of highways in Labrador. Phase I of the project was a $57 million upgrade of the Happy Valley-Goose Bay to Labrador City highway, which was completed in 1999. This section of highway was originally constructed in the early 1990s and was officially opened in mid 1992. Phase II included the construction of a $133 million gravel road from Red Bay to Cartwright, connecting the small communities along the southern coast of Labrador. Work was completed to Mary's Harbour in 2000, with a link to Cartwright anticipated in 2002. Construction of the remaining link between Cartwright and Happy Valley-Goose Bay would complete the Trans-Labrador Highway, but no schedule has yet been proposed.

CAUSEWAYS

Road construction over the rough Newfoundland and Labrador terrain invariably leads to the need for causeways. Many communities were located on islands, necessitating the use of ferries to reach the mainland. Where the separating distance was short, a causeway or bridge could be built. One of the earliest causeways was constructed in 1896 connecting the Conception Bay communities of Bay Roberts and Coley's Point. However, this was exceptional, as few causeways existed prior to Confederation. One of the earliest post-Confederation causeways was built on the Burin Peninsula in 1950, connecting Allan's Island with Lamaline. A more significant structure was the Hefferton causeway and bridge built in 1952 which connected Random Island in Trinity Bay with the mainland. In 1965, New World Island in Notre Dame Bay was connected via Chapel Island to the mainland with a 3030 foot causeway. In 1973, Twillingate was connected by causeway to New World Island, hence eliminating the ferry service to the latter and thereby ending two centuries of isolation. This causeway is the longest in the province. Over the past quarter century many less significant causeways have been built to make islands accessible or to shorten driving distances between points separated by water. Some of these are between Brighton and Cobbler Island, Brighton and Triton, Clarke's Head and George's Point, Pilley's Island and Robert's Arm, Pilley's Island and Triton Island, and Sunday Cove Island and the mainland.

Many causeways also incorporate bridges as part of their structure. Two such structures are part of the Trans-Labrador Highway; one spans the St. Lewis River, and the other, the Alexis River. The bridges at both locations are built of steel truss and steel grid decking using the static cantilever method of construction. With this technology, the bridge started on one side and was extended horizontally in a manner similar to the way tower cranes are extended vertically. Both of these bridges are 110 metres long, making them amongst the longest of their type in the world.

BRIDGES

One of the earliest recorded bridges in Newfoundland was King's Bridge in St. John's, which was built by the British Army's Ordnance Department in 1801. Undoubtedly there were earlier bridges spanning rivers and ravines, but they have been lost in the historic record. When organized road construction began in the mid 1820s, bridges were constructed of timber. It was not until the arrival of motor vehicles that it became apparent that timber would not be suitable, and either steel or reinforced concrete would be required. The first reinforced concrete bridge in Newfoundland was built in 1916 to span the Romaines River near Port au Port on the west coast. It consisted of two arched spans of seventy-five feet each. Yet it was not until 1924 that concrete became a standard construction material for new bridges. Ordinary concrete however had its limitations, but in the 1960s a new era in technology was introduced when pre-stressed concrete was first used in constructing a bridge in the province. It was relatively short, only a thirty-six foot span across Trout River near St. George's, and was completed in late summer 1960. The first major bridge using pre-stressed concrete was three spans of 75 feet each across Southwest Brook. Just about all new bridges on the island of Newfoundland now use pre-stressed concrete, whereas in Labrador, where the construction season is shorter, steel is more often used. The three-span Ossokmanuan Bridge on the Trans-Labrador Highway in Western Labrador uses a concrete deck slab with steel box girders. The bridge was designed to carry very heavy loads, such as 450 tonne transformers, which may at some time be needed at the Churchill Falls power plant.

> **North West River Cable Car**
>
> In 1961, a bridge of a different type went into service across the North West River in Labrador. A cable car was installed across the river enabling residents of the community of Northwest River to cross to the community of Sheshatshiu and access the highway to Happy Valley/Goose Bay. Towers 850 feet apart supported two cables, which operated a half ton car, capable of holding six passengers. The trip across the river took about 2 ½ minutes. This link was removed when a bridge was constructed in the late 1970s.

The government built many concrete arched bridges in the 1920s. An excellent example is the Cataracts Bridge, built in 1926 on the Old Placentia Road, six kilometres from Colinet. This is the oldest Newfoundland bridge which can carry vehicular traffic up to fifteen tonnes.

In 1928, the Grand Codroy bridge (335m) was built across the Grand River. This structure had eight spans and used the three hinged arch structural principle for the first time in Newfoundland. It remained the longest bridge in Newfoundland for fifty years until its collapse as a result of an ice buildup in 1978.

In the late 1940s, the government began installing a number of panel bridges (usually known as Bailey bridges). These were prefabricated structures based on designs used by the Allies during World War Two.

The bridges could be constructed, modified or removed in a very short time.

In 1961, a 320 foot single span steel bridge was erected across St. Paul's Inlet on the province's west coast. Construction of this structure posed technical difficulties because of the high tides and unpredictable

Bridge in Cataracts Provincial Park
Constructed in 1926, this is the oldest bridge capable of handling vehicular traffic up to fifteen tonnes.

winds in the area. Lundrigan Construction Limited assembled the bridge on the ground and hoisted it in place from a barge that was moored with ten thousand feet of cable to hold it against the tides.

The island's first highway overpass was built at Donovans, just west of St. John's, as part of the Trans-Canada Highway. A much larger 1220 foot structure was also built at Steady Brook, near Corner Brook; it has since been replaced. The Trans-Canada Highway also includes the province's first cloverleaf, which was constructed near Bishop's Falls. When officially opened in 1966, the highway required sixty-five bridges and nineteen overpasses.

> **Placentia Lift Bridge**
> One of Newfoundland's more interesting bridges is the Sir Ambrose Shea lift bridge across the Placentia gut. It was originally built in 1961 and comprises two approach spans of approximately thirty metres each, and a centre span of the same length. The centre section when raised provides approximately thirty metres clearance above the water. Construction design and project management for the structure was performed by Foundation of Canada Engineering Corporation Limited, and the main construction was performed by McNamara Construction.

Allandale Road Bridge
This Outer Ring Road overpass in St. John's is a typical pre-stressed concrete bridge. (Courtesy of the Department of Works, Services and Transportation)

One of the longest bridges in the province is the 224-metre-long structure across the Grand Codroy River, which was built in 1983. Newfoundland and Labrador's longest bridge is the 460 metre long CN viaduct in St. John's, constructed in 1979.

Some of Newfoundland's noteworthy highway bridges:

The ten longest fixed spans

CN Viaduct	St. John's	1979	460m
Grand Codroy River	Grand Codroy River	1983	224m
Sir Robert Bond	Exploits River	1958	192m
E. W. Winsor (steel)	Goose River	1967	177m
Northwest River	Northwest River	1981	172m
Gander Bay Causeway	Gander River	1968	168m
Corner Brook Harbour Arterial	Corner Brook	1974	162m
Junction Brook	Junction Brook	1965	160m
Searston Gut	Grand Codroy River	1991	153m
Ossokmanuan Reservoir	Ossokmanuan Reservoir	1992	145m

Moveable spans

Placentia lift-span (steel)	Placentia	1961	30m
Marystown (steel swing span) (no longer functioning)	Marystown	1959	20m

The last decade of the nineteenth century was a period of great bridge-building activity for the Newfoundland Railway, when more than 140 bridges and trestles were built. These were designed by the Dominion Bridge Company and were usually of steel construction. Over the years most of these bridges were replaced or upgraded, especially after Canadian National Railways took over the rail system in 1949. Since CNR decommissioned the railway and pulled out of Newfoundland in 1988, many of these bridges and trestles have been removed for safety reasons.

RAILROADS

The world's first steam powered locomotive made its appearance in Wales in 1804 when a steam engine was adapted to power a train hauling ten tons of coal. About twenty years later, in 1825, the world's first scheduled train service carrying passengers and freight was initiated between Stockton and Darlington in England. From that early start, rail transport expanded throughout Europe and the rest of the world. The emerging railroad systems in the nineteenth century were becoming important parts of the transportation infrastructure and the expansion of rail service became a high priority. In 1828, construction began on a line between Baltimore and Ellicott Mills, Ohio, the beginning of the famous Baltimore and Ohio Railroad. This line provided the first passenger train service in the United States. More than sixty years later, in May 1869, the first intercontinental rail line was completed when the eastern rail network was extended from Omaha, Nebraska to Sacramento, California.

> **St. John's Railway Station**
> The St. John's railway station was completed in late 1902 and opened on January 7, 1903. The building was of stone construction using granite quarried from the Gaff Topsails. The station was designed by W. H. Massey, who was chief engineer for the Reid Newfoundland Company, and built by Charles Henderson. On its first floor were the usual railway accommodations, including men's and women's waiting rooms, ticket office and baggage room. Upstairs were a number of offices, along with a huge bay window, where the station operator had a good view of the tracks.

In Canada, the first train service was inaugurated in 1836 when the Lake Champlain and St. Lawrence Railway opened for business. Other rail lines were quickly established to connect cities and towns in Ontario, Quebec, and the Maritimes. With the growth of rail service elsewhere in North America, it was not surprising that a train line across

A. L. Blackman steam engine – one of the first to serve the Newfoundland Railway (Courtesy of NHS)

Sir William Whiteway (1828-1908)
Whiteway was Prime Minister of Newfoundland 1885-1889, 1894, and 1897-1900. He was a strong supporter of the trans-island railroad. (Courtesy of PANL)

Newfoundland would also soon appear. After all, there had been a telegraph line across the colony since the mid 1850s. Surely a rail line, which would allow passengers and freight to board at St. John's and be transported to the west coast of the colony, would not be far behind.

During the late 1860s, construction of the Intercolonial Railway across the Canadian Maritime provinces was overseen by Sandford Fleming, a noted Canadian engineer of Scottish origin – another of whose innovations was a proposal to divide the world into the twenty-four standard time zones which are in use today. The construction of the Intercolonial Railway was one of the conditions of confederation between Nova Scotia, New Brunswick, Ontario and Quebec when they formed Canada in 1867. Later, Prime Minister Sir John A MacDonald, in an effort to establish a country from sea to sea, promised British Columbia a railroad connection to the east if they joined Canada, which they did in 1871. MacDonald followed through on his promise and on November 7, 1885, Donald Smith placed the last spike on the Canadian Pacific Railway, enabling train service from Montreal to the coast of British Columbia.

Fleming advocated the construction of an all-British sail and rail route connecting Britain with the Orient via Ireland, Newfoundland, and Canada. A railway across Newfoundland, a short ferry ride across the Cabot Strait, and a trans-continental train trip from Sydney, Nova Scotia to British Columbia would negate the need for Orient-bound passengers from Britain to take the long and perilous route around Cape Horn at the tip of South America. Fleming therefore had a personal interest in a railway across the island to St. John's. At his own expense, he sent a party of surveyors to Newfoundland to determine if it was feasible to build a rail line across the colony. Prior to this, little was recorded about Newfoundland's interior, apart from the reports of William Cormack, who had walked across the island in 1822, and the geological surveys (see Chapter Five) conducted by Joseph Jukes and Alexander Murray in the mid 1800s. Fleming's surveyors finished their work without incident and concluded that a railroad could be constructed across the island with no great difficulty.

TRANS-ISLAND RAILWAY

In 1874 Sir Frederick Carter, Prime Minister of Newfoundland, proposed to the Legislature that a survey be conducted for a rail route from St. John's to St. George's Bay on the west coast. The government contacted Sandford Fleming who agreed to undertake the project. Already familiar with the Newfoundland terrain from his earlier work, Fleming sent a team headed by A. L. Light, from the Intercolonial Railway, who along with government geologist Alexander Murray directed the survey. The surveyors and their assistants arrived in St. John's from Halifax on May 19, 1875 on the *SS Newfoundland*. They were divided into three teams, each assigned surveying responsibility for different parts of the route.

Fort William Train Station
This St. John's train station, which was part of a former English fort, burned down in 1900. (Courtesy of the City of St. John's Archives)

The survey extended from St. George's on the west coast to the head of Red Indian Lake, thence to the Gander River, Come by Chance, Long Harbour, Hodgewater, and then on to St. John's. Each team consisted of a survey engineer, transit man, and leveller. The teams steamed to their respective starting points – St. George's Bay, the mouth of the Exploits River, and Sunnyside, Trinity Bay. The survey began in June 1875 and was completed in December of the same year.

Robert G. Reid
Robert Gillespie Reid (1842-1908) was born in Coupar-Angus, Perthshire, Scotland, and emigrated to Australia in 1865 to prospect for gold. He was unsucccessful, and returned to Scotland in 1869. In 1871, he went to the United States, and eventually found himself in Ottawa, Canada, working on modifications to the Parliament Buildings as a mason, his original trade. He set up a construction company and obtained a number of contracts in the United States and Canada to build railroad bridges. This led to contracts for railway lines, one of the earliest being a line between Algoma and Sault Ste. Marie, Ontario. In 1889, Reid responded to a tender from the Newfoundland government to take over the construction of a railway from St. John's to the mining area of Halls Bay, after the original contractor defaulted. Reid became a wealthy businessman, and was knighted in 1907. He was a member of the Board of Directors of several organizations, including the Bank of Montreal, Royal Trust, and Canadian Pacific railways. He passed away in Montreal in 1908. (Courtesy of the A. C. Hunter Library, Newfoundland Collection)

It was estimated that it would cost about $8.5 million to construct a 360 mile-long standard gauge railway from St. John's to St. George's Bay, and in 1878 the government decided to proceed with construction. The project, however, required the ratification of the British government, which balked at the idea. The British Parliament refused to give its approval mainly because the railroad's proposed western terminus at St. George's was on the so-called "French Shore." This was a stretch of coast over which France controlled the fishery and restricted any construction within half a mile of the shoreline. Political considerations prevailed, and ultimately the plan for a railroad was quashed.

The plan for train service however would not go away. In 1878, Sir William Whiteway, a strong proponent of the trans-island railroad, became Prime Minister of Newfoundland. Whiteway and his government believed the interior of the colony should be made accessible for the development of mineral and forest resources, so he set up a committee to assess the viability of a railroad through the area. In 1880, the committee recommended construction of a railway from St. John's to Halls Bay (340 miles), connecting the towns along the route. The Halls Bay area was near the sites of copper mines at Tilt Cove, Betts Cove, and Little Bay. The legislature passed an Act authorizing funding (not exceeding $5 million, and $500,000 per annum) for construction of the railway. It also set up a five-man commission to oversee the project, consisting of R. Thorburn, C. R. Ayre, H. M. Mckay, and W. V. Whiteway, which was chaired by Ambrose Shea. The government requested bids for both construction and operation of the Halls Bay railway as well as for a spur from Whitbourne to Harbour Grace. The government decided to re-survey the trans-island route and commissioned Kinipple and Morris of London, England to do the work. On July 7, 1880, nineteen surveyors arrived on the *SS Nova Scotian* and began the survey.

The government received three bids for the proposed railroad. E. W. Plunkett and Associates submitted a proposal to build and operate a standard-gauge line, provided the government paid seventy-five per cent of the construction cost and granted five thousand acres of land for every mile of track laid. A consortium headed by lawyer Albert Blackman, including William Bond, Frank W. Allin, C. X. Hobbs, and Domingo Vasquez, proposed a three foot-six inch narrow-gauge line, requiring a government subsidy of $250,000 annually, as well as five thousand acres of land for each mile of track laid. The third submission was from J. N. Greene and Associates, a construction firm from Saint John, New

Brunswick, which proposed a three-foot gauge railway requiring a government subsidy of $246,000 annually, as well as one thousand acres of land for each mile of track constructed.

Of the three submissions, the government favoured the Blackman proposal, and awarded the consortium the contract. The agreement covered a thirty-five-year period and called for the company to clear the land, lay the track, provide the rolling stock and also operate the railway. The final contract required the government to provide the consortium a subsidy of $180,000 per annum; however the group had to post a $100,000 US bond. At the expiry of the contract the government would have the option of buying the railway.

Railway and steamer routes circa 1933 (Courtesy of the MUN Map Library)

The Blackman consortium began the job in August 1881. Construction was by pick and shovel with horse and cart. By January of the following year, about twenty miles of rail line had been completed west from Fort William, at the eastern end of St. John's. By year end, the line had reached Holyrood, and by the following summer, the Conception Bay North area. The first steam locomotive, which was purchased from the Prince Edward Island Railway, arrived in St. John's on December 5, 1881 aboard the *SS Merlin*. Regular train service began to Topsail in June, 1882 and to Holyrood later that year.

The Blackman group, however, began experiencing financial difficulties, and started cutting corners (quite literally, by not meeting curve specifications). By the spring of 1884, the company could no longer continue construction and declared bankruptcy. The bondholders, however, decided to continue the project, and by late 1884, had completed the line as far as Harbour Grace. In November of that year, regular train service commenced between that town and St. John's.

In the fall of 1885, Robert Thorburn, who was earlier a member of the commission overseeing railroad construction, defeated Whiteway's government and became Prime Minister. His government stopped construction of the Halls Bay railroad line, but gave the go-ahead to build a twenty-six-mile long spur from Whitbourne to Placentia. This line cost $465,000 and was completed in 1888.

The political scene changed again in 1889 when Whiteway's Liberal party defeated Thorburn's Tories in an election and Whiteway once more became Prime Minister. He immediately took up the challenge to complete the railroad he had initiated nine years earlier and finish it all the way to Halls Bay. The government felt that hiring a contractor to finish the railroad would be expensive, so it decided to proceed on its own and recommence construction from Placentia Junction. Finding that the cost of railroad construction was higher than it anticipated, the government also decided to re-visit its decision and look at other alternatives. It subsequently decided to request quotations to finish the line; this eventually led to discussions with Robert G. Reid (1842- 1908) and G. H. Middleton, both of whom had considerable experience in building railroads in Canada and the United States. By 1890, Reid and Middleton reached an agreement with the government to complete the railway for $15,600 per mile. About two years after construction recommenced, Middleton left the partnership and Reid took over the project on his own.

After mining activity in the Halls Bay area began to subside in the 1890s, the government decided to change the railway's western terminus, and instead place track across the Gaff Topsails and make Port aux Basques the western end of the rail line. At Port aux Basques, a scheduled passenger service to Sydney in Nova Scotia would be set up, enabling train passengers and freight to conveniently connect with the ferry. The government and Reid negotiated a new contract for this change, with terms similar to the one already in place. Later, Reid bargained for a revision which would see him operate the trans-island and Placentia railways for a ten-year period, and also receive five thousand acres of land for every mile of rail road constructed. All told, the new agreement would add approximately 285 miles of railway line, extending from the mouth of the Exploits River across the Gaff Topsails to Port aux Basques. By the spring of 1898, Reid had completed the line across the island. The first passenger train departed St. John's on June 29 and arrived at Port aux Basques more than twenty-eight hours later.

In 1898, Whiteway's government was defeated by the Conservative party under James S. Winter. Winter's government again renegotiated the agreement with the Reids, giving them the right to operate the railway for fifty years, as well as an additional five thousand acres of land per mile of rail built. Reid's heirs were also to receive $1,000,000 at the end of the contract.

A scandal erupted when it was revealed that Alfred B. Morine, one of Winter's senior cabinet ministers, had helped renegotiate the contract while he was also retained as one of Reid's lawyers. In the election of 1901, Winter was defeated by Robert Bond, who once again renegotiated the deal with the Reids. Up to this point, the railway contract had been with Reid himself, but for the 1901 contract, Reid incorporated the Reid Newfoundland Company Limited, with a capitalization of $25 million. Robert Reid remained head of the company up until his passing on June 3, 1908. His son William. D. Reid (1867-1924) succeeded him as President of the company.

The first railway station in St. John's was at Fort William, near the site of the present day Fairmont Hotel and Fort William Buildings. The building was a concrete structure, originally part of the British fortifications at Fort William, which were decommissioned in the 1870s. It was destroyed by fire in March of 1900 and subsequently, a new station was erected at the western end of Water Street. The new structure was constructed from stone quarried on the Gaff Topsails, and became operational in 1902. The building still stands, and is currently being prepared to house the City of St. John's Archives, a move which is scheduled for late 2002.

SPUR ROUTES

After completion of the line to Port aux Basques, the Newfoundland government decided to continue construction and extend the railway to other areas. Additional railroad construction was probably justified on its own merit, but was likely also an election ploy on the part of the Whiteway government to re-employ the many labourers laid off after completion of the trans-island line. The government subsequently contracted with the Reid Company for construction of three spur lines – from Clarke's Beach to Tilton, from Harbour Grace to Carbonear, and from Notre Dame Junction to Lewisporte.

In 1909, the government further contracted with the Reid Company for additional railway track to link other communities. The first of these was a branch line between Clarenville and Bonavista which was completed in 1911. In 1913, new tracks were placed and train service was implemented between St. John's and Trepassey, followed the next year by a spur to Heart's Content, and a line between Carbonear and Bay de Verde in 1915. For these lines, the Reid Company was paid $15,000 per mile and granted 4,000 acres of land for every mile of track constructed.

During this era, the Reid Company was not the only concern constructing railroads. In 1909, the Anglo-Newfoundland Development Corporation (AND) built a twenty-two-mile-long rail line between Grand Falls and Botwood. This line served several purposes including transporting paper from the Grand Falls mill to the Botwood seaport, bringing coal and other supplies to Grand Falls, and transporting concentrate from the Buchans mine to the port. The line also provided passenger service between Grand Falls, Bishop's Falls and Botwood. The so-called 'Botwood Railway' was reorganized in 1957 and became a common carrier named the Grand Falls Central Railway. In the winter months, when Botwood was usually icebound, paper from the Grand Falls Mill was shipped via rail to either Heart's Content or St. John's, ports which were normally open.

NEWFOUNDLAND RAILWAY

The Reid Company operated the Newfoundland Railway until the early 1920s. In 1920, the company advised the government it was incurring cash flow problems and could no longer run the railway. To manage the rail service, the government provided operating funds and set up a commission consisting of appointed officials and representatives of

The St. John's Dry Dock
In 1883, the government decided that a dry dock was necessary to provide maintenance for the increased shipping in the colony. It contracted with J. E. Simpson and Company of New York, which completed the dock in December 1884. The $550,000 dry dock was about six hundred feet long, one of the largest such facilities on this side of the Atlantic. The government leased the dry dock to the Simpson Company for $15,000 per year; however, by 1892, the company had gone bankrupt, and the government was forced to operate it on its own. In 1894, the government found a new operator and leased the facility to Angel and Harvey Company for $11,000 per year. In 1898, the Reid Company purchased the dry dock for $325,000. The photograph above shows the dry dock under reconstruction in 1926. (Courtesy of the City of St. John's Archives)

the Reid Company. This arrangement continued until the end of June 1923, when under the "Railway Settlement Act" the government bought out the Reid Company's contract. The government paid the company $2,000,000 in government bonds, and the Reids maintained ownership of the approximately six thousand square miles of land they had acquired over their thirty-three year involvement with railroads in Newfoundland. The Reid Company sold its ownership in the St. John's Street Car Company, the St. John's dry dock, its electrical distribution and generating facilities, and some of its land.

After the Reid Company bowed out, the government set up the Newfoundland Government Railway to run the rail service, and in 1923, appointed Herbert Russel to head the operation. The company's name was shortened to Newfoundland Railways in 1926.

In one of its early ventures, Newfoundland Railways constructed a line from Millertown Junction to Buchans in 1927 to provide service to the mining operation there. Its operations continued with various route changes over the years, including the shutdown of the Heart's Content, Bay de Verde, and Trepassey routes in 1931. During World War Two, the line between St. John's and Argentia was heavily used by the United States Military for transport of construction workers and material for the American base. Because of the considerable pressure it exerted on Newfoundland Railway's tracks, the American military upgraded the rail line between Placentia Junction and Argentia, and also contributed approximately one hundred flatcars, passenger cars, and other rolling stock. When Newfoundland became part of Canada in 1949, the Newfoundland Railway was taken over by Canadian National Railways (CNR). At the same time, CNR also assumed responsibility for the St. John's dry dock, the Port aux Basques to Sydney ferry service, the coastal service to the outport communities, the Newfoundland Hotel, and the telegraph and telephone services provided by the Department of Posts and Telegraphs.

CANADIAN NATIONAL RAILWAYS

The railway system taken over by CNR in 1949 was below standard in many respects. The track was old, and many of the bridges and culverts needed replacement. The track itself had numerous excessive grades and many curves were too tight. CNR considered converting the narrow gauge (forty-two inch) track to a standard gauge (fifty-six and one half inch); however it elected to continue with the existing track, although massive improvements would be required. CNR replaced 306 miles of track with heavier iron, installed almost 2.5 million new ties, and replaced or repaired many culverts and bridges. Between 1953 and 1959, the steam locomotives were replaced by fifty-six new diesel units and new freight cars were added, bringing the rolling stock count to more than twenty-one hundred. New rail yards, freight sheds, and shops were built at St. John's, Bishop's Falls, Corner Brook and Port aux Basques. By the end of 1966, CNR had spent approximately $55 million upgrading the Newfoundland railway.

A velocipede on the rail line near Steady Brook (Courtesy of the Department of Works, Services and Transportation)

Since CNR's operations in the rest of Canada used standard gauge, the use of narrow gauge track in Newfoundland caused problems for the company. For instance, the rolling stock entering or leaving Newfoundland had to have the wheels changed to suit the appropriate gauge. CNR tackled the narrow gauge issue with ingenuity and success. In addition to being one of the first to modify standard gauge freight cars for use on narrow gauge, the company also designed and built, in its St. John's yards, the world's first bi-level automobile freight cars.

The running of the railway in Newfoundland was a costly proposition, which became more pronounced when the Trans-Canada Highway was completed across the province in 1965. The completion of the highway resulted in a decline in railway use for both passengers and freight. The journey by highway from St. John's to Port aux Basques took only about twelve hours, whereas the train trip required almost twenty-four hours. In the late 1960s, CNR announced that it was reducing its passenger train routes, and in 1968 was given permission by the government to introduce "Roadcruiser," a trans-island bus service, which began operation in December of that year. On July 2 1969, the last passenger train from St. John's to Port aux Basques, affectionately called the "Bullet," pulled out of the St. John's station. CNR continued passenger service on some routes, but began to focus its operation on the transport of freight. By 1984, Terra Transport, a division of CNR that operated the railway, had proceeded with the closing of thirty-two train stations in the province. After a Royal Commission into Newfoundland's transportation system, and negotiations between the provincial and federal governments, a deal was reached in 1988 whereby the railroads in Newfoundland would close down completely and the province would be compensated $640 million toward the upgrade of its highway system. On September 30, 1988 the final train ran from Corner Brook to Bishop's Falls, and within a couple of years almost all the rail track had been removed. Only ninety years

SS Bruce
*Built in 1897 for the Reid Company, the eleven-hundred-ton **SS Bruce** was the first ferry to provide scheduled service between Port aux Basques and Sydney (Courtesy of the A.C. Hunter Library, Newfoundland Collection)*

after the first train trip between St. John's and Port aux Basques, the railroads had disappeared from the insular part of the province, to become but a footnote in Newfoundland's history.

OTHER NEWFOUNDLAND RAIL SYSTEMS

Apart from the lines built by the Reid Company, the Newfoundland government and CNR, the only other significant rail activity on the island involved the paper and mining companies, which used railroads to deliver their product to a port or to connect with the Newfoundland Railways tracks. Apart from the spur from Grand Falls to Botwood and the rail operations in Labrador, none of these railroads carried passengers.

As of 2002, the only railroad remaining in the province is the Quebec North Shore and Labrador Railway (QNS&L), which was built between Sept-Iles and Schefferville, Quebec in 1954. The line crosses a portion of Labrador, and its main purpose is the transport of iron ore from the mines in Quebec and western Labrador to the seaport at Sept-Iles. A fifty-eight kilometre spur route was constructed in 1960 from Ross Bay Junction to connect Labrador City to the main line. QNS&L also provides a limited passenger service, the only train passenger service in the province.

MARINE SERVICES

From the discovery of Newfoundland and Labrador by Europeans in about 1000 AD, up until the advent of modern highways and airline service, travel both to and within the region has been by sea; yet organized "ferry" operations did not begin until the 1800s. The early passenger ships were not designed for the regular transport of people, and travellers obtained rides at the whim of the ships' captains. The first recorded instance of a commercial ferry service in Newfoundland appeared in an 1808 entry in the *Royal Gazette*. The paper makes mention of a Michael Dooley, who operated out of Portugal Cove, conveying passengers around Conception Bay. In the early 1800s, a number of "packet" services arose in the St. John's, Conception and Trinity Bay areas, and later, other services appeared in other parts of the island.

The first regular passenger and mail service between St. John's and Halifax was provided by the Allan line via the *SS North American*. Service began in June 1844, when the ship arrived in St. John's after a sixty hour voyage. This steamer service replaced an earlier sailing packet service which had been provided by Bland & Tobin.

In order to boost trade, the Canadian government began subsidizing ferry service between Canada and Newfoundland as early as 1897. Steamship service was inaugurated between Halifax and Bonne Bay with points en route; between Halifax, St. John's, and Liverpool, England; and between Prince Edward Island, Cape Breton Island and eastern Newfoundland.

The Reid Company began a ferry service between Placentia and Nova Scotia with the *SS Bruce* in October 1897. After the railway was completed to Port aux Basques in June, 1898, the *Bruce* was moved to Port aux Basques for passenger service to North Sydney. The ship served this route until it sank in 1911 near Louisbourg in Nova Scotia, after which it was replaced by the *Glencoe* and the *Invermore*. The Canadian government began subsidizing the gulf run in 1906 and continued this practice until the mid 1920s. After the Newfoundland government took over the ferry service, only one ship serviced the route.

In accordance with its 1898 contract, the Reid Company was required to provide and operate ferry service in the colony. There were to be eight ships in the fleet: one for the Labrador service, one for the Cabot Strait, and six for coastal service around the island. At the time the government was already operating seven ships: the *Grand Lake* and *Virginia Lake*, for the south and north coasts of the island; the *Leopard* for the Labrador run; the *Bruce* for the gulf; and the *Lady Glover*, *Favourite*, and *Alert* for the Notre Dame, Trinity, and Placentia Bay areas. Since better ships were required to improve service, the Reid Company began to procure additional ferries. The new ships were constructed in Scotland, and all were christened by Robert Reid's sister. The so called "alphabet fleet" consisted of the *Argyle* (Placentia Bay run), *Bruce* (Cabot Strait), *Clyde* (Notre Dame Bay), *Dundee* (Bonavista Bay run), *Ethie* (Trinity Bay service), *Fife* (Labrador), *Glencoe* (Placentia to North Sydney route and the South coast run), and *Home* (Humber mouth to Battle Harbour). Over the years, some of these ships were lost at sea or replaced by others.

In 1904, the Newfoundland government, in an effort to improve transportation links along the coast, entered into a contract with Bowring Brothers to augment the service provided by the Reid Company. Bowrings operated the *Prospero* on the north coast and the *Portia* on the south until the contract expired in 1924.

During World War One, the Russian government was in dire need of ships with ice-breaking capability to operate in its northern ports. In 1916, the Reid Company accepted an offer from Russia for $750,000 to purchase the second *Bruce* (built in 1911) and the *Lintrose*. Once the *Bruce* was sold, the *Kyle* began operation on the Gulf run.

The Newfoundland government took over Reid's dockyard, railway, and coastal services in 1923. The ships acquired included the *Argyle*, *Clyde*, *Glencoe*, *Home*, *Kyle*, *Meigle*, *Petrel*, and *Sagona*.

During World War Two, the Newfoundland government built the "splinter fleet" at its shipyard in Clarenville. These wooden ships, named the *Bonne Bay*, *Burin*, *Clarenville*, *Codroy*, *Exploits*, *Ferryland*, *Glenwood*, *Placentia*, *Trepassey*, and *Twillingate*, were operated by the Newfoundland Railway. After Confederation, CNR took over the Newfoundland ferry operation. The ships in service at the time were the *Baccalieu*, *Bar Haven*, *Burgeo*, *Cabot Strait*, *Glencoe*, *Kyle*, *Northern Ranger*, and *Springdale*. Many ships were dropped from or added to the fleet over the years.

In 1925, the *Caribou* replaced the *Kyle* on the gulf run and remained the main link with Canada for the next seventeen years. The impact of World War Two on Newfoundland struck home on October 14, 1942, when the *Caribou* was sunk by a German U-boat, with the loss of 137 lives.

The *MV William Carson* began freight service between Argentia and North Sydney in 1955. The 7500 ton *Carson* was built by Canadian Vickers and had a capacity of 500 passengers and 110 vehicles. When the new port facilities at Port aux Basques were completed in 1958, the *Carson* began passenger service between that town and North Sydney. The *Carson* was joined on the Port aux Basques-North Sydney route by the *Patrick Morris* in 1965 and the *Leif Eriksson* in 1966. In 1967, summer passenger ferry service was also implemented between Sydney and Argentia, when the *Ambrose Shea* began service. The *Carson* stayed on this run until 1976, when it was replaced and transferred to the Lewisporte-Goose Bay service. Near Battle Harbour on June 2, 1977, the *Carson* hit an ice floe, which cut a hole in its hull. All passengers and crew safely boarded lifeboats, but the ship sank in about three hours.

CNR began receiving delivery of new ships for the coastal service in 1956, after which some of the older ships were decommissioned. New arrivals to the fleet included the *Bonavista*, *Hopedale*, *Nonia*, *Petite Forte*, and *Taverner*.

In 1984, CN Marine was established to handle CNR's marine operations in the province. In 1987, it split completely from CNR and became Marine Atlantic, a federal crown corporation. In 2002, Marine Atlantic's service was limited to two routes: between Port aux Basques and Cape Breton and between Argentia and Cape Breton. The company's ships include the *Clara and Joseph Smallwood* and its sister ship the *Caribou,* the largest ferries in Canada. Other Marine Atlantic ships operating on the gulf service in 2002 were the *Leif Ericson* and the *Atlantic Freighter.*

Just about all other ferry services in the province are operated by the provincial government. In 2002, some of these routes include Portugal Cove to Bell Island (*Flanders* and *Beaumont Hamel)*, St. Barbe to Blanc Sablon (*Apollo)*, Fogo Island to Farewell (*M/V Captain Earl W. Winsor)*, Lewisporte to Goose Bay (*M/V Sir Robert Bond*), and St. Anthony to Nain (*M/V Northern Ranger*).

NC-4
On May 16, 1919, the NC-4 US Navy seaplane left Trepassey, and landing via the Azores and Lisbon, Portugal, became the first aircraft to fly the Atlantic. (Courtesy of Larry Sheehy)

AVIATION

Newfoundland's geographic position as North America's closest landfall to Europe has given it historical significance for many transatlantic flights. In 1913, the London Daily Mail offered a £10,000 prize for the first non-stop transatlantic flight between Newfoundland, Canada, or the United States and the United Kingdom. The contest was late getting started as it was suspended during World War One. In 1919, Lieutenants J.C.P. Woods and C.C. Wylie attempted to fly from Kent, England to America, but crashed in the Irish sea, becoming transatlantic aviation's first victims. This was the first of many flights by aviators that were to meet an ignominious end.

The first successful crossing of the Atlantic by air was by a fleet of US Navy Curtis flying boats. This project employed approximately sixty support ships along the flight route in case the planes came down in the sea. The plan was to fly from Trepassey harbour to the Azores and then on to Plymouth, England. On May 16, 1919, three flying boats designated the NC-1, NC-3, and NC-4 took off from Trepassey. The first two planes were forced down within a few hundred miles from the Azores; however, the crews were recovered. The NC-4, under the command of

John Alcock (1892-1919) and Arthur Whitten Brown (1886-1948)
Alcock on the left and Brown on the right were the first aviators to fly the Atlantic non-stop, departing from St. John's, Newfoundland June 14, 1919; they landed in Clifden, Ireland seventeen hours later. (Courtesy of MUN CNS Archives)

Lieutenant Commander A. C. Read, continued on and landed at Horta, the Azores, seventeen hours after departing Newfoundland. On May 27, 1919, Read flew the NC-4 on to Lisbon, Portugal, and four days later continued to Plymouth. NC-4's transatlantic flight was obviously not non-stop, so did not qualify for the Daily Mail prize. The flight, however, was recognized as the first to cross the Atlantic.

Several weeks after the Americans completed their crossing, John Alcock (1892-1919) and Arthur Whitten Brown (1886-1948) attempted the first non-stop transatlantic flight. They left Lester's Field (in what is now central St. John's) on June 14, 1919 in a Vickers Vimy biplane, powered by two Rolls-Royce 350 horsepower engines, and crash-landed about seventeen hours later in a bog at Clifden, Ireland. They won the Daily Mail's £10,000 prize for a non-stop Atlantic crossing and became worldwide celebrities. Both Alcock and Brown were knighted for their achievement. Alcock, however, had little time to enjoy his fame as he was killed in an air crash late in 1919. It was not until 1927 that the first solo non-stop Atlantic flight was achieved. This was accomplished by Charles Lindbergh, who on May 20 took off from New York and headed toward St. John's. Although not landing in Newfoundland, he used Signal Hill as

Amelia Earhart (1898-1937)
Earhart left Trepassey in June 1928 to become the first female to fly the Atlantic as a passenger. In May 1932 she took off from Harbour Grace and also became the first woman to fly the Atlantic solo. (Courtesy of MUN CNS Archives)

his last sight of land before heading towards Europe, landing at Le Bourget airport in Paris, France.

In June 1928, Amelia Earhart (1898-1937) left Trepassey as a passenger, to become the first female to cross the Atlantic by air. On May 20, 1932, Earhart made history again, when at the controls of a Lockheed Vega Monoplane, she took off from the Harbour Grace airfield to begin the first solo transatlantic flight by a woman. She landed at Londonderry, Ireland early the next day.

Newfoundland also played an important role in other transatlantic flights in the decades to follow. In the summer of 2003, two gallant airmen plan to fly the Atlantic in an antique Vickers Vimy aircraft to recreate Alcock and Brown's 1919 transatlantic achievement. In the history of transatlantic flight, Newfoundland's location has given it an important role and continues to make it the scene of many significant firsts.

BOTWOOD SEAPLANE BASE

Newfoundland's geographic position continued to be a major factor in both civil and military aviation until the 1980s. In the 1920s, Sidney D. Cotton, an Australian, and Sydney Bennett, a Newfoundland associate, used Botwood as a base for their mail delivery and seal spotting services. The site also became popular as a refuelling stop for transatlantic flights. In 1935, Botwood was visited by British officials who recommended it as refuelling base for the "flying boats" which were being planned for transatlantic passenger service. These aircraft were huge by the standard of the time, with multi levels providing sleeping accommodation and eat-

The Caledonia
These "flying boats" provided transatlantic service via Botwood between 1937 and the early 1940s. (Courtesy of PANL)

Famous Visitors
In its short history a long list of dignitaries passed through the Botwood seaplane base. Some prominent politicians were Winston Churchill, Lord Mountbatten, and Lord Beaverbrook. Famous entertainers included Douglas Fairbanks Jr., Gracie Fields, Edgar Bergen, Edward G. Robinson, and Bob Hope.

ing facilities. The first experimental flying boat landed in Botwood on 5 July 1937 when the Pan American Clipper III (a Sikorsky-42 aircraft) arrived from Shediac, New Brunswick. The next day, on July 6, it departed for Foynes, Ireland, en route to Southampton, England. On the same day, a flying boat left Foynes, and landed at Botwood at 7:36 a.m. British Overseas Airway Corporation (BOAC) and Pan American Airways soon began regular mail and passenger service across the Atlantic. The flying boat service continued for a couple of years, but was interrupted when war broke out in 1939. In 1941, operation of the Botwood base was turned over for the duration of the war to Canada, which operated a fleet of Canso reconnaissance aircraft from the site. With the construction of a new airport at Gander to provide the refuelling needs of transatlantic aircraft, the importance of the seaplane base at Botwood diminished and it eventually closed down.

GANDER INTERNATIONAL AIRPORT

In the mid 1930s, the British and Canadian governments began looking at potential landing sites in Newfoundland for transatlantic land-based flights. After a preliminary search, surveyors determined that a site near Gander Lake would be ideal for a transatlantic airport. The location was convenient because the lake could be used by flying boats in times of emergency or when Botwood was not available. Construction of the airport began in 1936, and by the end of 1938 the runways had been completed. Much of the navigation equipment used at the seaplane base in Botwood was transferred to Gander, and the new airport began operations.

In 1940, the Royal Canadian Air Force established a military base at Gander. During World War Two, RCAF aircraft provided reconnaissance and convoy support from the base which was defended by Hurricane fighters. The base also served as a critical refuelling stop for airplanes manufactured in North America which were being ferried to England. The Canadian military controlled Gander airport during the war, but after the conflict, the Newfoundland government resumed authority.

After the war, Gander airport became a popular refuelling site for transatlantic air carriers, including Pan American, BOAC, and Trans World. Trans-Canada Air Lines added Gander to its schedule in 1942. The airport was such a busy spot for transatlantic flights that it was nicknamed "Crossroads of the World." After Confederation in 1949, the Canadian government took over the airport and set upon a major expansion and improvement program to allow Gander to attract increased air traffic. By the 1950s, Gander airport was handling more than one thousand passengers a day, and was serving eight international airlines. The airport continued as a transatlantic refuelling stop until the 1990s, when new generations of long-range jet aircraft eliminated the need for refuelling on most transatlantic flights. However, the airport is still used for domestic airline service, as well as the occasional international refuelling.

Transatlantic Robotic Airplane

*On August 20, 1998, the **Laima**, a miniature robotic airplane, was launched from Bell Island and headed towards Europe. The craft weighed only 13.2 kilograms fully loaded and carried one and a half gallons of fuel. The **Laima** was built by the Insitu Group, a Washington, U.S. company specializing in robotics, with financial support from a number of different agencies. The aircraft was guided by satellite and landed at the Hebrides Islands twenty-six hours later.*

WARTIME AIRPORT CONSTRUCTION

Newfoundland Airports (as of 2001), along with their longest runway

Black Tickle 2500 ft
Cartwright 2500 ft
Charlottetown 2500 ft
Churchill Falls. 5500 ft
Davis Inlet 2500 ft
Deer Lake 6000 ft
Gander 10500 ft
Goose Bay 11000 ft
Hopedale 2500 ft
Makkovik 2500 ft
Marys Harbour 2500 ft
Nain 2000 ft
Port Hope Simpson 2500 ft
Postville. 2500 ft
Rigolet 2500 ft
St Anthony 3900 ft
St John's 8500 ft
Stephenville 10000 ft
Wabush 6000 ft
Williams Harbour 2000 ft

World War Two brought about a flurry of airport activity in Newfoundland. In 1941, the Royal Canadian Air Force began construction of airports at Goose Bay, Labrador and Torbay (now St. John's International Airport). In May, 1942, Trans Canada Airlines began one-flight-a-day service between St. John's and mainland Canada, using a ten passenger Lockheed Lodestar, which connected to various mainland cities. Air service revolutionized travel between Newfoundland and the Canadian mainland by reducing travel time from a matter of days via train and ferry to just a few hours by air.

The United States military forces also built airports at Stephenville and Argentia, and established an Air Force Base at Goose Bay in Labrador, where they maintained their Strategic Air Command. On June 30, 1976, Transport Canada took over the operation of the civilian part of the Goose Bay airport. The airport currently serves as a Canadian Armed Forces Base which is used by the air forces of several NATO countries for practice of low-level flying. More detail on military construction during World War Two is given in Chapter Seven.

AIRPORTS IN 2002

There are currently twenty airports in Newfoundland providing scheduled passenger service. The airports capable of handling jet aircraft are located at Churchill Falls, Deer Lake, Gander, Goose Bay, St. John's, Stephenville, and Wabush. Thirteen other smaller airports scattered throughout the province provide propellor airplane service. There are also several communities which have airstrips for emergency or recreational aviation.

Zanussi
Newfoundland's involvement with transatlantic flights did not end with Amelia Earhart. In 1978, two British balloonists, Donald Cameron and Chris Davey, lifted off from St. John's in their Zanussi balloon. About one day east of the city, a tear in the balloon occurred, causing it to lose altitude. They continued on, but on the fifth day, the wind subsided and they ditched in the Atlantic, only one hundred miles from the coast of France.

Tilt Cove circa 1873
Tilt Cove was the the site of Newfoundland's first significant mine.

Chapter Five

Mining - From Ballast to Bullion

A simple definition of mining is the systematic removal of rocks or minerals from their original source for a special use. In Newfoundland, five thousand years ago, the Maritime-Archaic Indians used hard stones to break up Ramah chert (a hard grey rock which flakes easily, found near Ramah Bay in northern Labrador) in order to fashion it into spearheads, knives and other tools. About sixteen hundred years ago, the Dorset Eskimo began quarrying soapstone near Fleur de Lys on the island's Baie Verte peninsula. They used the soft soapstone to create utensils such as cooking pots and oil lamps, as well as for jewellery. Later, the Beothucks discovered rhyolite, from which they constructed knives, spearheads, and other implements. Newfoundland's aboriginal peoples had not learnt how to extract minerals such as copper or iron; mining for metals such as these did not begin until the arrival of Europeans, who sought minerals for economic gain. Mining as an industrial endeavour will be the focus of this chapter, and Newfoundland's history in this regard is as deep and rich as the mineral deposits that have been exploited.

References to Newfoundland's minerals occur early in its written history. Sir Martin Frobisher made mention of a "shiny heavy stone" found

*Alexander Murray (1810-1884)
Murray was the first director of the Geological Survey of Newfoundland. (Courtesy of the A. C. Hunter Library, Newfoundland Collection)*

in what is assumed to be Trinity Bay in 1576. In 1578, Anthony Parkhurst, a British explorer, returned to England with samples of copper and iron ore, retrieved from around St. John's and Bell Island. Sir Humphrey Gilbert, on his famous voyage of 1583 to claim Newfoundland for Britain, brought along a "mineralogist" who retrieved copper, iron, lead and silver ores from the Avalon Peninsula. Other early visitors to Newfoundland were primarily interested in the fishery; however the lure of gold, silver and other minerals was never far from their minds.

Probably the earliest non-aboriginal mineral operation in Newfoundland was a copper mine at Shoal Bay, just south of St. John's on the Atlantic coast. This was begun by several local individuals who operated it until the late 1770s. However, the mine did not prove to be profitable and closed down after a couple of years.

Prior to the nineteenth century, Newfoundland was largely regarded for its fish resources and the British government discouraged permanent settlement. In the early decades of the nineteenth century however, the local population grew considerably through immigration from southwest England and southeast Ireland. The mainstay occupation of the settlers was fishing, and few had either the time or the inclination to explore the land for minerals.

THE GEOLOGICAL SURVEY OF NEWFOUNDLAND

In the early nineteenth century, geological surveys were routinely conducted by industrialized countries to search for mineral deposits to help advance industrial development. In the late 1830s, the Newfoundland government decided to follow this route and hired an English geologist named Joseph Jukes to perform a geological survey of the island. Jukes explored the island in 1839 and 1840; however, because of lack of funding, the government curtailed his work, so his report provided little significant information. It was not until 1864 that the government decided to recommence the survey. It hired Alexander Murray (1810-1884), a Scottish geologist, to oversee the first official geological survey of Newfoundland. Murray was a former assistant to Sir William Logan, who was head of the geological survey of Canada.

James Howley (1847-1918)
Howley was the director of Newfoundland's Geological Survey between 1883 and 1909. (A. C. Hunter Library, Newfoundland Collection)

Murray became Newfoundland's first geological survey director and organized the first systematic examination of the island's geology. He was joined by James Patrick Howley (1847-1918), who became his assistant in 1869. Murray resigned in 1883 and Howley became director, a position he held until 1909. These men added greatly to the knowledge of the island's geology, and under their leadership published many maps indicating new waterways, forestation, agricultural and mining opportunities.

After Howley's retirement, the geological survey was suspended. However, Princeton University had a strong interest in Newfoundland's geology and sent many expeditions to explore for minerals. In 1926, the Newfoundland Geological Survey recommenced and H. A. Baker served as director until 1929.

A. K. Snelgrove was the next government geologist, who was appointed after the Commission of Government was established in 1934. Snelgrove served until 1943 when C. K. Howse became director. After Confederation in 1949, the position was eliminated and future geological

Mines Operating 1855-1860 - *adapted from F. N. Gisborne (From Once Upon a Mine, by Wendy Martin)*

Harbour Mille, Fortune Bay	Copper and silver
Turk's Head, Conception Bay	Peacock copper
English Ridge, Conception Bay	Grey copper
La Manche, Placentia Bay	Galena or lead
Frenchman's Hill,	Peacock copper
Griffin's Point	Peacock copper
Sweetman's Island	Silver lead
Strouter, Placentia Bay	Silver lead
Rockey Cove	Grey copper
Stoney House Cove	Grey copper
Lawn	Silver lead
Paquet, French Shore	Yellow copper
Terra Nova, Little Bay	Mundic and copper

surveys were undertaken under the auspices of various government departments and programs related to mining and resource development. In the 1970s, geological survey activity was re-activated. Since then, the Newfoundland Geological Survey has had a number of directors, including Frank Blackwood P. Geo., who was head in 2002. The Survey is currently a section of the Newfoundland Department of Mines and Energy and has a staff of approximately forty-five professional and support personnel.

EARLY MINING

The first entrepreneur to take a serious interest in mining was Charles Bennett (1793-1883). Bennett was born in Dorset and came to Newfoundland in the early 1800s. He became a wealthy St. John's businessman, whose interests included a brewery, bank, shipyard, sawmill, and foundry. During the mid 1800s he was a member of the House of Assembly, serving as Prime Minister from 1870 to 1874. His interest in mining began in the early 1850s when he obtained the mineral rights to one million acres of land in the Fortune and Placentia Bay areas. This grant was recalled in 1856 when it was cancelled by the Liberal government because of Bennett's vocal opposition to the government's favourable position towards Responsible Government. He continued however to explore for minerals, and ultimately was permitted to keep ten five-mile-square claims.

Another personality in nineteenth century mining was Frederick Newton Gisborne, who is better known for his role in telegraph communications in the province (see Chapter Two). After serving as chief engineer for the New York, Newfoundland and London Telegraph Company during the construction of the 1856 trans-island telegraph line, Gisborne began to focus his interest on exploring for minerals, having come across mineral deposits in his work on the project. During the survey of the La Manche area of Placentia Bay, a lead deposit was discovered, for which the New York, Newfoundland and London Telegraph Company staked a claim. Cyrus Field was a major shareholder in this company, and in 1857, a mining operation was set up by his brother Matthew Field and other New York investors, including a Major Ripley and a Mr. Crockett. Gisborne, however, was not directly involved in this mining operation.

Charles Fox Bennett (1793-1883)
An early mining promoter who also served as Newfoundland's Prime Minister 1870-1874 (Courtesy of PANL)

In 1857, the La Manche operation hired Harry T. Verran, an expert in mining from Cornwall, England, who would become the first mining engineer to work in Newfoundland. His involvement in this concern was short-lived. After about a year, a dispute with the owners over unsafe work practices arose and he resigned. Despite his departure, the mine was worked on and off for ten years, but was never a profitable operation.

In the late 1850s, Gisborne explored for minerals in the Placentia, Exploits, Bonavista, and Notre Dame Bay areas. He entered into a partnership with Bennett and opened copper ore mines at Turk's Head and English Ridge, Conception Bay. His contribution to Newfoundland's mining history was not limited to prospecting and promotion, as he also represented the Newfoundland Mining Association (established in 1857). This organization was set up by English merchants with the objective of encouraging mining operations in Newfoundland. After the failure of the Conception Bay mining venture, Gisborne decided to abandon mining in Newfoundland and left the province to pursue other ventures in mining and telegraphy. He returned again around 1860 to re-open a mine; however it once again failed.

THE MINES
TILT COVE

Tilt Cove, located on the west side of Notre Dame Bay, was the location of the first significant mine in Newfoundland. Copper was discovered there in 1857 by Smith McKay, a prospector, who in partnership with Charles Bennett formed the Union Mining Company and began mining operations in 1864. The company brought in experienced miners from Cornwall, England, and also hired local Newfoundlanders. The ore was blasted and cobbed at the surface and transported by rail to the dock, where it was loaded aboard ships for Wales. By the end of the 1860s, the mine employed almost three hundred men, with almost eight hundred residents living at the town site. Bennett bought out McKay's interest in 1880, but died three years later. In 1886, a new mine was opened east of the existing operation, as the ore from the original mine was near depletion. In 1888, Bennett's trustees leased the property to the Tilt Cove Copper Company Ltd. of London, England, which built a smelter at the site to extract the copper ore. The new mine

> *The wreck of the Queen of Swansea*
> *A poignant story involving Tilt Cove involved the wreck of the **Queen of Swansea**. The 360 ton ship was used primarily as a ferry and small freighter and on December 12, 1867, was carrying passengers from St. John's to Tilt Cove, when a snow storm came up, crashing the ship on the rocks of Gull Island, not far from Tilt Cove. Four men drowned while trying to salvage provisions. Eleven survivors were stranded on Gull Island with no food, water or shelter.*
>
> *Several of the survivors kept written notes or wrote letters for their families, the latest letter dated December 24, 1867. All perished and it was not until four months later that their bodies were discovered by bird hunters who had visited the island.*

also yielded small quantities of gold and silver. The smelter proved uneconomical and closed in 1890.

Market conditions for copper, however, later improved and production increased. By 1901, the population of Tilt Cove had swelled to almost fourteen hundred. The mine operated sporadically up to the end of World War One when the ore could no longer be economically recovered, and mining operations ceased. Up to that point, the estimated output of copper from the mine was about sixty-one thousand tons.

In 1957, the Maritime Mining Corporation reopened the mine and discovered new bodies of copper and gold ore in the area. The ore was depleted in 1967, and the mines again shut down after shipping 467,760 tons of copper concentrate.

BETTS COVE

One of the more interesting mines in Notre Dame Bay from a historical point of view was at Betts Cove. The mine was about eight miles south of Tilt Cove, and opened in 1875. The man behind the Betts Cove mine was none other than a German baron, Francis von Ellershausen (1830-1914). Ellershausen was a mining engineer who arrived in Nova Scotia in 1862 to work in a gold mining operation. After a couple of years, he left engineering and entered the pulp and paper business, amassing a small fortune. A mining engineer friend, Adolph Guzman, was expected to marry his daughter, but she broke off the relationship and in 1872 Guzman fled to Newfoundland where he began prospecting in the Notre Dame area. For unknown reasons, he staked claims in the Betts Cove area in Baron Ellershausen's name. Guzman continued prospecting in Newfoundland, and left the country in 1883 for the United States, where he was reportedly murdered.

During the 1870s, Ellershausen headed to Newfoundland to organize the mining operations. He subsequently sailed to Britain, and along with William Dickson and Walter Mackenzie, two Scottish capitalists, set up the Betts Cove Mining Company (BCMC). He hired thirty German miners and brought them to Betts Cove, who along with seventy Newfoundlanders began copper mining in 1875. BCMC constructed Newfoundland's very first smelters at Betts Cove in 1876. The blast furnaces, which were fuelled by coal from Swansea, Wales, reduced the copper ore to a twenty to thirty percent concentrate. In the first three years of operation more than seventy-five thousand tons of ore were shipped to

Swansea. By 1878, the population of the mining town swelled, the majority Newfoundlanders, but with many miners from Germany, Canada, England, and as far away as Australia. In 1880, Dickson, one of the original BCMC shareholders, passed away. Dickson's death appears to have greatly affected Ellershausen and he decided to get out of the Betts Cove mining business. In December 1880, he sold his assets in BCMC for $2.5 million to the Newfoundland Consolidated Copper Mining Company (Consolidated). Consolidated was not as fastidious as Ellershausen in operating the mine, and cut corners by poor mining practices. These shortcuts resulted in part of the mine collapsing in 1883, and by 1886 the last load of ore was shipped to Swansea. Over the life of the mine, 130,692 tons of ore had been shipped, as well as 2450 tons of pyrite.

As with most one-industry mining towns, Betts Cove became a ghost town almost overnight. During its peak in the late 1870s, it had a population of around 2000, but by 1901, there were only thirteen residents.

LITTLE BAY

The copper deposit at Little Bay, Notre Dame Bay was discovered by a local resident, who subsequently sold his find to holders of min-

Little Bay Mine
The site of one of Newfoundland's early copper mines (Courtesy of the NHS)

ing rights in the area. These included Adolph Guzman, who turned the property over to the Betts Cove Mining Company. The discovery of the Little Bay deposit coincided with the winding down of the Betts Cove Mine, so many of the latter's buildings, equipment, and employee accommodations were floated down the bay to Little Bay about thirty kilometres to the southwest. The new mine site started production in late 1878 and in its first year of operation shipped approximately ten thousand tons of ore to Swansea for processing.

In 1880, the Newfoundland Consolidated Copper Mining Company took over the operation, and three years later, the layoff of one hundred men led to a short strike. In 1887, smelters were installed which produced copper ingots from the ore, and helped improve the mine's profitability.

The Little Bay town site grew and by 1884 had more than fifteen hundred inhabitants, increasing to more than twenty-one hundred in 1891. The life of most mining towns is precarious. By 1894, inefficient operating methods coupled with low copper prices forced the mine to close, and compelled the workers to move to other mines in Newfoundland and Canada.

In 1955, the New Highridge Mining Company surveyed the property and estimated a reserve of 2.6 million tonnes of 2.6% grade copper. By the early 1960s, worldwide demand for copper had increased, allowing the mine to be re-opened. In May 1961, the Atlantic Coast Copper Corporation started operations and shipped copper concentrate to a mill at Murdochville, Quebec, which recovered almost fifty-seven million pounds of copper. The mine closed down on October 31, 1969.

During the copper boom at Tilt Cove, Little Bay, and Betts Cove, other locations in the Baie Verte and Notre Dame Bay area were explored for minerals, and substantial copper finds were made at several places. A number of other copper mines opened in the area, including Terra Nova (1850s) and Consolidated Rambler at Baie Verte (1964), Pilley's Island (1887), Whalesback (1965), and Gullpond (1967).

BELL ISLAND

Rocks from the iron ore deposits on Bell Island in Conception Bay had been used for decades before the commercial value of the iron they contained was realized in the late 1800s. The red ore was much heavier than common sandstone, and for years, fishermen had used the mate-

The conveyor belt loading ore at Scotia Pier Bell Island (Courtesy of MUN CNS Archives)

rial for killicks and ballast. As early as the 1620s, John Guy's men from Cupids noticed Bell Island's hematite ore, and had some sent to England for analysis; results of their tests are not known.

It was not until 1892 that a Mr. Butler from Port de Grave noted that the iron in the rocks on Bell Island might have commercial possibilities and sent some to Canada to be assayed. The tests were positive, and the following year the rights to the ore deposit were obtained by the New Glasgow Iron, Coal and Rail Company of Nova Scotia, which later through a merger became the Nova Scotia Steel and Coal Company (Scotia). Overseeing the mine was R. E. Chambers, the company's chief engineer. The company began mining operations in 1895.

The first mine on Bell Island was an open pit operation, with ore dug by manual labour. It was located on the northwest side of the island, but the ocean near the island at this location was too shallow to allow boats to take on ore. The company therefore built a trestle pier on the southeast side of the island where the water was deep enough for large ore boats. A dual track tramway with cable-driven ore cars was built to transport the ore approximately two miles from the mine site to the pier. One track was used for carrying the ore to the pier and the other as the return path for empty cars. The open pit mine lasted until ore ran out around 1909.

Mining engineers determined that there were three main veins of iron at the Bell Island site. In 1899, Scotia sold the rights to the upper and lower beds of the deposit to the Dominion Iron and Steel Company (Dominion), and kept the higher grade middle bed for its own use. The upper bed was approximately eight to sixteen feet thick, the middle sixteen feet thick, and the lower ten to forty feet thick. Dominion built a new pier a few hundred yards north of the "Scotia" pier, as well as a tramway across the island to their mine site. By 1901, approximately eleven hundred workers were employed by the two companies.

In 1902, surface ore from the lower and middle beds ran out, and the companies began sinking shafts underground. The miners travelled down to the mine face in empty iron ore tram cars which were lowered and raised by a cable. The Bell Island mines were of a "room and pillar" type. During the excavation of the iron veins, about fifty to sixty per cent of the ore had to remain behind to support the structure. On the night shifts, blasters used dynamite to loosen the ore. The morning shift saw the "face cleaners" going into the mine to ensure no loose ore was hanging overhead. They were followed by the "muckers," or shovellers, who were expected to load by hand twenty tons of ore per shift in the dark and damp mine, illuminated only by candles or oil lamps.(Carbide lamps were introduced around 1911 and battery lights came on the scene in the 1930s.) Horses were used to haul the ore carts from the mine face to the central tram-track. They were rested in stables deep within the mines where they spent their entire working lives, rarely seeing daylight before they were retired because of age. From the central track, the ore carts were joined together and pulled up by a steam-driven cable winch.

Stanley J. Carew (1914-1977) Carew was a Bell Island engineer who started with the mining company in 1938 and left there in 1941 to become head of engineering at Memorial College. He became Dean of Applied Science in 1949, a position he maintained until becoming master of Paton College in 1967. He served as Deputy to the President and Director of Conferences at Memorial University from 1972 until his passing in 1977. (Courtesy of MUN Photographic Services)

Work horses hauling iron ore up the slope of a Bell Island mine

The mine shafts extended six kilometres out beneath Conception Bay. Over the course of the mines, the shafts were interconnected, providing additional safety routes in case of emergency.

Prior to World War One, Europe was a major market for Scotia's ore. With the outbreak of hostilities, the market evaporated, and Scotia ceased mining operations. Dominion closed down one of its mines and laid off fifteen hundred workers. After the war, the economic recession put the damper on any immediate market recovery.

British investors bought a majority of Scotia's and Dominion's shares and merged the two companies in 1921 to form the British Empire Steel Corporation (Besco). Besco did not have mining management expertise and managed the operations poorly. As a result, it was taken over in 1926 by its mortgager, the National Trust Company, which ran the Wabana mine until 1930. At that time, the Dominion Steel and Coal Company of Nova Scotia (DOSCO) acquired the mining operations under its subsidiary Dominion Wabana Ore Limited.

> **World War Two Attacks**
> On September 5, 1952, World War Two came to Bell Island when the first two of four ore ships were sunk by German torpedoes. The **Lord Strathcona** *(with no loss of life), and the* **Saganaga** *(29 lives lost), both loaded with iron ore and anchored off Little Bell Island, were waiting to join a convoy when they were attacked and sunk by a German submarine. On November 2, another submarine attack sank the ore carriers* **Rose Castle** *(28 lives lost), and the* **PLM 27** *(12 lives lost), while they were anchored off Lance Cove. Scotia Pier was also hit by a torpedo which had missed its target.*

During the depression of the 1930s, there were limited markets for Wabana iron ore. For periods of time, some of the mines were either closed or operated on a reduced workweek. The one bright spot was Germany. In the late 1930s, the demand for iron from Germany was so high that all four mines were operating at full production. That came to an end in

The conveyor belt on Bell Island extended from the minehead, approximately two miles across the island to Scotia Pier. Originally the ore was transported by rail cars, which were replaced by Euclid 20-ton trucks in the late 1940s. The belt for the "rubber railway" was manufactured by the B. F. Goodrich Company, and was the longest such conveyor in the world. (Courtesy of MUN CNS Archives)

September of 1939 when German tanks invaded Poland triggering World War Two, with tanks in all likelihood constructed from steel made from Bell Island iron. Iron ore production continued during the war, with Wabana now feeding raw material to steel plants in Britain and Nova Scotia for the production of war supplies.

After the war, ore production increased, as steel was needed for European post-war rebuilding projects. To help capitalize on this opportunity, DOSCO embarked on a $20 million modernization program and appointed W. L. Stuewe, an engineer from its Sydney office, as the mine's manager. In the 1950s, new mechanical equipment was installed, including Joy units (each consisting of a loader, two shuttle cars and one or two drill mobiles), electric shovels, and ventilation equipment. The remaining horses were then retired. Modernization activities also included the removal of the rail tracks between the mines and the pier, and their replacement by twenty-ton Euclid trucks to carry the ore across the

> **Mining Casualties**
> *As with any mining operation dating from the nineteenth century, it was to be expected that the hazardous working conditions on Bell Island would cause many casualties, including deaths. Safety standards were non-existent in the early days, but did improve over time, yet not before the mining operations on Bell Island had claimed more than one hundred lives, the first occurring in 1898 and the last in 1965, shortly before the closing of the mine. Most accidents involved only one or two men, and thankfully there was no major catastrophe such as a mine collapse. The largest single calamity was a dynamite explosion in 1916 which killed four men.*

island. The trucks were eventually superseded with a conveyor belt system which carried ore up from the mines and across the island to Scotia Pier, where it was loaded aboard ships for export. This conveyor system was the longest system of its type in the world.

In the early 1950s, steel plants began to convert to a low cost process which allowed them to produce excellent steel from high-grade, low phosphorous iron such as that from Labrador, Australia and elsewhere.

Scotia Pier, Bell Island (Courtesy of PANL)

Frederick W. Angel M.B.E. (1874-1937)
Angel, who graduated from McGill University in 1898, was one of the early Newfoundland-born engineers. He worked with the Reid Company, Wabana Mines, and the Oliver Mining Company in Minnesota, before joining his father's firm in St. John's at United Nail and Foundry, where he served as president from 1913 to 1938. During World War One, one of his companies produced two thousand tons of shell casings for the war effort. (Courtesy of Roger Angel)

Wabana ore had 0.8% phosphorous content, more than double the amount suitable for the new mills, so its ore lost out to the new finds. In spite of DOSCO's modernization program, the high phosphorous content of Wabana's ore sounded the death knell for the mine, and made markets scarce.

In 1957, DOSCO's operations were taken over by A. V. Roe (Canada) Limited, which in 1962 became Hawker Siddley Canada Limited. Ore production decreased and layoffs began in 1959 when Number 6 mine closed down, followed by the closure of Number 4 mine in 1962. Mining continued on the island until 1966 when Number 3 mine also closed, bringing Bell Island's long mining history to an end. The mines could not compete with the newer iron mines in Minnesota and Labrador which produced a higher quality ore at a lower cost as well. DOSCO's land and mining equipment were sold to the Newfoundland government.

When mining on Bell Island was at its peak in the 1950s, Wabana was the second largest community in the province, with a population of around twelve thousand. The island had a number of amenities, including an ice arena, tennis courts, movie theatres, a curling rink, and the best track and field sports complex in the province. In 2002, thirty-six years after the mines closed, approximately twenty-seven hundred people still live on Bell Island, by which which currently is the scene of very little industrial activity. While a few of the residents work with service industries on the island, most are either unemployed or commute to work in St. John's. A bright light for the island was the recent conversion of part of old Number 2 mine into a mining museum, which now attracts thousands of visitors annually.

BUCHANS

On January 7, 1905, Alfred and Harold Harmsworth, two English brothers who were publishers of London's *Daily Mail* and *Daily Mirror* respectively, incorporated the Anglo-Newfoundland Development Company. AND was primarily interested in the production of paper and had obtained a concession on cutting rights in central Newfoundland's forests to feed its proposed mill at Grand Falls. Its concession also included a right to develop minerals in the area. The company hired William Canning, who had studied mining engineering at Montreal's McGill University, and Mathieu (Mattie) Michel, a local Montagnais-Micmaq, to conduct a survey of the area. One of their objectives was to explore for sulphur, which was a necessary mineral for the manufacture of paper. In the fall of 1905, they explored the north shore of Red Indian Lake near the Buchans River, and Michel pointed out deposits of lead, zinc, and copper that he had earlier discovered. They found no sulphur, but AND felt that the lead and copper find might have commercial prospects. In the following year the company began an exploratory mine. William Scott, an AND engineer who had earlier worked with the Reid Company, oversaw the operation and hired a few experienced miners from Cornwall as well as miners from Notre Dame Bay to delineate the deposit. He would eventually become AND's Chief Engineer, and later General Manager and Vice-President. AND transferred its mining and mineral operations to a subsidiary, Terra Nova Properties Limited, in 1908.

Drillers at the Buchans Mine (Courtesy of the Red Indian Lake Development Association)

In 1910, one thousand tons of ore were shipped to Sweden for testing. The results were discouraging because it was determined that the complexity of the ore made it impossible to extract the metals economically by conventional methods of the time, and to do so, a costly electrical smelter would be required. This news discouraged the company and it

Mathieu Michel, also known as Mattie Mitchell, the discoverer of the original Buchans ore outcrop. In 2002, the federal government named him a person of national historic significance. (Courtesy of the Red Indian Lake Development Association)

put its mining ambitions in the Buchans area on hold. Several years later, the American Smelting and Re-fining Company (ASARCO), a large metals company headquartered in New York, became aware of AND's disappointing test results. ASARCO felt that its engineers could develop an economical extraction method, so it contacted AND regarding obtaining mining rights and samples of ore from the area. Twenty-five pounds of ore from the mine were sent to ASARCO's laboratory in the United States, where different experiments on extracting lead and zinc were conducted. In 1925, the lab succeeded in developing an economical procedure using a selection flotation technique. With this new process, ASARCO decided that the ore at Buchans could be profitably mined. The following year, it negotiated a twenty-five year deal whereby it would receive the mining rights within a twenty mile radius of the Buchans mine, and AND, in return, would receive fifty per cent of the mining profits derived from the area.

In June 1926, ASARCO hired the Swedish American Electrical Prospecting Company, led by Hans T. Lundberg, to explore the Buchans area using a new electrical prospecting technique. Lundberg, a past Professor of Mining at the Royal Technical Institute of Stockholm, was the first to use employ electrical geophysical prospecting in Newfoundland. After several

The crew which discovered the "Lucky Strike" find, with Hans Lundberg on the far right (Courtesy of George Neary, P. Eng.)

The Lucky Strike Mine with the Buchans townsite in the background (Courtesy of the Red Indian Lake Development Association)

weeks, Lundberg and his team discovered two significant ore bodies – the so called "Oriental" and "Lucky Strike." After determining the economic viability of these finds, ASARCO set up the Buchans Mining Company (Buchans Mining) and began discussions with AND to renegotiate their agreement. The new agreement increased the term to fifty years, and extended ASARCO's mining rights to a radius of thirty miles (from the original twenty miles) from the Buchans mine.

The Buchans Mining Company began developing a full scale mine in late 1926. Since there was a railway stop at the nearby town of Millertown, it was used as the staging camp for the accumulation of construction materials, equipment, food and other supplies. The material was transported over the frozen lake to the mine site by tractor and sled. At the site, a crushing plant was built, as well as a concentrating mill and storage sheds. The construction crew numbered up to nine hundred workers, who completed the concentration plant and Lucky Strike buildings by mid 1928.

The mine required a transportation link to get the concentrate to market, as well as electric power, both of which were provided by AND. A twenty-two mile long railroad (the Buchans Railway) was built which connected into the Newfoundland Railway at Millertown Junction. This enabled Buchans ore to be delivered to the seaport at Botwood entirely by rail. For electric power, the Buchans River was dammed at Sandy Lake and a six foot diameter penstock was constructed to feed a 1.865 megawatt hydro-electric plant one mile downstream. Electric power was first produced in January 1928. Disaster struck on July 30 when the dam collapsed, severely damaging the powerhouse and destroying two sections of railroad trestle. After repairs were made, power was restored, and the mill began processing its first ore on September 1, 1928.

To accommodate expansions to the milling operation, additional electric power would be required. This was obtained from the International Power and Paper Company's Deer Lake power station which was connected to the Buchans site via a forty-eight mile long power line installed during the summer of 1931.

In 1935, Buchans Mining also began to extract copper from its ore and opened up the Oriental ore body. Over the years, new areas were discovered and developed – the Rothermere (1947), MacLean (1950), Oriental No.2 (1953), and Clementine (1960).

At the Buchans town site in 1928, there were about sixty houses, a post office, churches and a hospital. All houses had electricity, water and sewage systems, making the town one of the most progressive in Newfoundland. Over the next few years, the Buchans town site expanded, and the usual buildings and services of a small town developed. ASARCO, however, would not permit private ownership of homes; rather the miners and their families were housed in company-owned buildings. By 1960, the town (under control of ASARCO) had a hockey rink, curling club, swimming pool, bowling alley, and a tennis court. Conditions further improved in 1963, when the provincial government laid out a new town site adjacent to the company-controlled town, allowing residents to build and own their own homes. It was not until 1956 that Buchans was connected to the Newfoundland highway system, when a highway was built between the town and Badger. The town was incorporated in 1979.

Buchans' mining operations were very successful, with ore shipped to Britain, the United States and Europe. Mining output was strongest in the 1930s, when the mines produced, on average, more than four hundred thousand tons of ore per year. This dropped to between three and four hundred thousand tons over the next three decades. In the 1970s, dwindling reserves reduced output to about two hundred thousand tons annually.

Rothermere mine at Buchans (Courtesy of the A. C. Hunter Library, Newfoundland Collection)

In 1961, Price Company Limited (Price) purchased AND's interests in Newfoundland, thereby becoming a partner in the mining operation. In 1976, Price obtained 51% of the operation, making ASARCO a junior partner. By 1983, the Buchans mine was no longer economic to operate and it closed down in 1983. Since its opening in 1927, approximately 17.5 million tons of ore had been recovered, yielding 3.8 million tons of zinc, 1.8 million tons of lead, 0.58 million tons of copper, as well as small amounts of gold and silver.

ST. LAWRENCE

The existence of fluorspar in the St. Lawrence area of the southeast Burin Peninsula had been known since the arrival of early European settlers. The colourful veins of fluorspar ore near the town were initially of interest because of the small amounts of lead and silver they contained. The quantity of these metals however was uneconomical for mining purposes and it was not until the early twentieth century that fluorspar became valuable in its own right as an ingredient for the manufacture of steel and aluminum.

In the late 1920s, prospectors descended on St. Lawrence to stake claims on the many veins of fluorspar in the area. One such prospector was John Taylor of St. John's. After filing claims for several areas, he went to New York where he sold two of his claims to Walter E. Seibert. Seibert was an accountant who had earlier been in St. John's on business and had heard of the St. Lawrence fluorspar deposits. He bought additional claims over the next year and acquired more than three dozen in all. Seibert set up the St. Lawrence Corporation of Newfoundland Limited (the Corporation) in 1931 to start mining the fluorspar. With the Great Depression coming on, Seibert did not have access to much cash. His company made an arrangement with a local St. Lawrence resident,

St. Lawrence fluorspar mine (Courtesy of APEGN)

Aubrey Farrell, whereby it would provide the machinery, if Farrell would provide the men to mine ore from an open pit on one of its claims. The critical proviso was that the miners would only be paid after two thousand tons of ore had been sold. The machinery provided by Seibert was used equipment that he had purchased for only $2,500 in a bankruptcy sale. It arrived in the winter of 1933 and was dragged across the ice to the Black Duck vein. Mining began, and by the spring of 1934, the requisite two thousand tons of ore had been extracted and delivered to DOSCO at Sydney. Seibert met his part of the deal and the workmen were paid.

Full commercial mining began the same year, and by the late 1930s fluorspar was being mined from several veins in the area. The Corporation operated on what was virtually a shoestring budget until the outbreak of World War Two, when the manufacture of steel for the Allied war machine guaranteed sure markets for its fluorspar.

In the 1950s, St. Lawrence saw increased competition from mines in Mexico, which produced ore which was not only cheaper to recover, but also of a higher grade. 1957 was a bad year for the Corporation. A major contract with the US government expired and there were no new markets in sight. The Corporation also received a setback when its Blue Beach mine caved in. To further complicate matters, a provincial government inquiry into the Corporation's affairs revealed that Seibert had a financial interest in the Mexican mines, the Corporation's main competition. The company struggled after this revelation and by 1961 was effectively bankrupt. Shortly thereafter it sold its equipment in St. Lawrence and departed from the province.

Other mining interests at St. Lawrence, however, continued some years beyond this. In May 1937, Edwin Lavino, president of E. J. Lavino and Company, a Philadelphia metals firm, set up the American Newfoundland Fluorspar Company (ANFC) and began a

---

St. Lawrence and Industrial Disease

Drilling for fluospar created a tremendous amount of dust, which was always of concern to miners. However, it was not until the 1950s that the first case of silicosis was confirmed at St. Lawrence, although the local union had been complaining to government officials about the problem since the late 1930s. Many former miners were diagnosed as having tuberculosis, which in many cases was an incorrect diagnosis. A further alarm was raised when many miners contracted various types of cancer, which was later traced to radon gas that was released during the mining process. In 1969, a Royal Commission examined the St. Lawrence situation, and made recommendations regarding safety procedures and compensation.

fluorspar mine near the claims of the St. Lawrence Corporation. ANFC had purchased its claims from Hookey and Company of St. John's which had staked them earlier in the 1930s. In late 1939, ANFC sold its rights to the St. Lawrence area to the Aluminum Company of Canada (Alcan), which incorporated a new company, Newfoundland Fluorspar Limited (NewFluor) to manage its investment. As a subsidiary of the huge Alcan company, NewFluor had the financial strength to ensure its mine was properly equipped and its workers well paid. This was in contrast to the operations of the Corporation, where working conditions and wages were poor. NewFluor had a ready market for its ore, as it was purchased by Alcan for use in its aluminum smelter at Arvida, Quebec. Newfluor also acquired the Corporation's abandoned property after it had left the province, and re-opened part of the mine.

In the early to mid 1970s there were several labour interruptions. Alcan still required fluorspar for their Arvida mill, and during the St. Lawrence labour disputes, purchased ore from Mexico. Alcan found that the high grade and inexpensive Mexican ore suited their aluminum operation and decided to continue its use in their mill. This was the end of the line for the miners of St. Lawrence, and Alcan filed notice to close down their operation effective February 1, 1978.

The people of St. Lawrence however had a much greater problem than the closing of the mines. Up until the 1940s, St. Lawrence miners used a dry drilling technique to get at the ore. This drilling method generated a considerable amount of silica dust. Over a period of time the very fine dust particles damaged their lungs, causing scar tissue and severely incapacitating breathing. The result was a condition called silicosis, which was often fatal.

Silicosis was not officially confirmed until the late 1950s, but by then hundreds of miners had been affected. By the early 1960s, it was noticed that there was also a high incidence of cancer in the St. Lawrence area, specifically among miners. An investigation into the problem eventually identified the probable cause as radon gas, which was detected in poorly ventilated areas of the mines. The ventilation systems were eventually improved, but far too late for many of the miners.

LABRADOR CITY

What appeared to be iron ore was first described by Europeans in 1870 in the writings of Louis Babel, a Swiss Oblate missionary who had travelled extensively in western Labrador. Its physical existence was confirmed in a mapping done by Albert Peter Low, a geologist with the Geological Survey of Canada in the early 1890s. The remoteness of the ore discouraged any great commercial interest at the time. It was not until 1936 that Weaver Minerals Limited of Montreal took an interest in the area and obtained prospecting rights. This company was later reconstituted as the Labrador Mining and Exploration Company Limited (Labrador Mining). Hollinger North Shore Exploration Company bought controlling interest in the company in 1942.

The company began a detailed geological survey of the western Labrador area, under the direction of Dr. Joseph Retty (1891-1945). This survey resulted in the discovery of iron deposits with great economic potential. The problem of course was that the deposits were located hundreds of miles in the interior. The cost of building the infrastructure as well as almost 600 kilometres of rail line to the nearest port would be enormous. To make such a large development viable, the Hollinger Company merged in 1949 with several US companies (ARMCO, Labrador Mining, National, Republic, Wheeling and Youngstown) to form the Iron Ore Company of Canada (IOC). Some of the owner companies were steel manufacturers who would purchase some of the iron ore. A group of Canadian and American insurance companies also helped to finance the deal and with financing and markets secure, IOC was ready to begin development.

> **Labrador City**
>
> *The pelletizing plant in Labrador City has a capacity of approximately 12.5 million tonnes per year and is the largest in Canada.*
>
> *In the Labrador City area, the iron ore reserve is approximately 4.1 billion tonnes.*
>
> *Labrador City is the largest town in Labrador with a population of more than 9000.*

IOC's first mine was at Knob Lake, Quebec, near the Labrador border. Construction of Knob Lake's infrastructure began in 1950. The major components of the project were a 573 kilometre railway from Sept-Iles in Quebec to the mine site; the hydro-electric facility at Menihek; the mining facilities, including processing plants, roads, and accommodation; and the port facilities at Sept-Iles. The construction of the town site at Schefferville (on the Quebec-Labrador border), the mine, and the railway required a massive airlift into the western Labrador and northern Quebec areas. The aircraft used lakes and ponds as well as temporary airstrips cut

Boxcars of pelletized iron ore awaiting shipment by rail to Sept-Iles, Quebec (Courtesy of the Iron Ore Company of Canada)

out of the forest. More than 80,000 tons of freight were transported and approximately 169,000 passenger flights were made in what many consider the largest civilian airlift up to that point in time.

The Quebec North Shore and Labrador Railway (QNS&L) between the mine area and Sept-Iles was completed in 1954, and the first rail shipment of iron ore departed on July 15. The mine at Knob Lake was closed down in 1982 and as of 2002, most of the town of Schefferville has been abandoned.

In 1958, IOC began setting up mining operations in the Carol Lake, Labrador area and the following year started construction of the Labrador City town site. The Carol Lake mine also presented huge engineering challenges. The mine site was about sixty-five kilometres from

Labrador City (Courtesy of the Iron Ore Company of Canada)

the QNS&L railway, and as with IOC's earlier construction in Quebec, personnel and equipment were airlifted using temporary landing facilities. Luckily for IOC, Wabush Mines was working an exploratory mine just a few kilometres away from Carol Lake, and both companies agreed to share the cost of constructing a spur to connect with QNS&L's main line. In 1960, the link was completed. Once the mines began production, the Carol Lake ore was shipped to Sept-Iles, and the Wabush ore to Pointe Noire, both in Quebec.

There was also no electric power in the area. Again the two mining companies joined forces and partnered with the Hamilton Falls Power Corporation to set up the Twin Falls Power Corporation, which constructed a hydro-generating facility at Twin Falls as well as a 110 mile long transmission line to the two mine sites. The first phase of the Twin Falls installation was a 89,500 kilowatt unit which went on stream in May 1962. After IOC's decision to construct a pelletizing plant at Labrador City, the output of the Twin Falls plant was doubled to 179 megawatts. The project was managed by Shawinigan Engineering Company and the general contractor was Dufresne Engineering Company. The transmission line was contracted to Canadian Hoosier Engineering Company.

Meanwhile work continued at the Carol Lake mine and its town site, which was named Labrador City. By 1961, a school had been built, followed shortly after by shopping, recreational and medical facilities. By 1965 the growing town also had a hospital, newspaper, and television station. IOC engineers decided to construct a concentrating and pelletizing plant at the site. The mine and concentrator were officially opened by Premier Joseph R. Smallwood in 1962 and the pelletizing plant came on stream in April 1963. In 1966, a magnetic separation process was added to the complex, allowing a higher grade concentrate to be obtained from the ore. February 28, 1989, was a milestone for IOC when it extracted its one billionth ton of iron ore from the Labrador City mine.

Since 1986, the IOC has also been mining dolomite, a mineral used in the production of fluxed pellets. In 1992, IOC opened a quarry at Leila Wynne, eighteen kilometres north of the pelletizing plant. In 2001, the quarry produced approximately eight-five thousand tonnes of dolomite.

IOC plant at Labrador City (Courtesy of the Iron Ore company of Canada)

In April 1997, North Limited purchased controlling interest in IOC and began a $650 million improvement program for its Labrador City facilities. In August 2000, the Rio Tinto Company, a large international mining conglomerate, purchased North Limited, obtaining a 56.1% ownership in IOC. The Mitsubishi Corporation owns 25% of IOC, and the Labrador Iron Ore Income Fund the remaining 18.9%.

The Labrador City mine has proven reserves of approximately 1.5 billion tonnes as well as an inferred reserve of up to 3.9 billion tonnes. In 2000, IOC produced 16.2 million tonnes of concentrate and pellets and employed more than fifteen hundred employees in its Labrador City mining, mill and pellet plant operations.

WABUSH

The mining rights to the area south of Wabush Lake were originally held by the Labrador Mining and Exploration Company, a predecessor company of IOC. The company felt that the area could not be profitably mined, and in 1953 it relinquished its rights to the government. These were eventually obtained by the Canadian Javelin Company, which was controlled by John C. Doyle. Doyle established Wabush Mines Limited the following year and began looking for a company to operate the mine. The successful enterprise was Pickands Mather and Company (Pickands) from Cleveland, Ohio. Pickands leased a five square mile concession and after surveying the area, estimated the deposit contained more than one billion tons of 36% grade iron ore. The company built a pilot mill at Wabush which produced one hundred thousand tons of 68% iron concentrate during 1961 and 1962. With these good results, Pickands set up a joint venture with several large steel and iron companies to begin full-scale production. These included Canadian companies (Steel Company of Canada Limited (Stelco) and Dominion Foundries and Steel Limited (Dofasco)); American companies (Youngstown Sheet and Tube Company, Inland Steel Company, Pittsburgh Steel Company, and Interlake Steel Corporation); German companies (Hoesch A. G. and Mannesmann A. G.); and the Italian company Finsider.

The consortium raised $235 million in funding and began construction of the mining facilities. The main mining area was named the Scully mine, after William Scully, a former chairman of Stelco. The ore was mined from an open pit, the dimensions of which were about one by three and a half miles. The ore was blasted and loaded by electric shovels into trucks for transport to the mill where it was crushed into football-size chunks in one of two gyratory crushers. It was then separated into coarse and fine sizes and moved via conveyors to a storage building. From there it was fed into the milling building, where water was added and the ore was ground in one of several autogenous grinding mills to a consistency approaching that of beach sand. The grinding mills were turned by 1750 horsepower electric motors.

The iron was next separated from the waste using centrifugal force, with the tailings pumped to the waste area and the concentrated iron pumped to the dryer building. There the concentrate was further processed by removing silica with electrostatic separators. The concentrate, consisting of 66% iron and 2.5% silica, was then stored in three 1000 ton silos, and later loaded into ninety-ton tank-type rail cars for shipment the 275 miles to Pointe Noire where it was pelletized before shipping to market. Modifications to the spirals in the concentrator in 1999 increased the quality of the product, resulting in an increase in output.

The Wabush town site was constructed across from the mine on the other side of Jean Lake. By mid 1965, when the mine was at full production, there were more than three hundred families in the town, living

Wabush mine site (Courtesy of Wabush Mines)

in modern, attractive housing units, while the single workers lived in residence halls. The fifty-room Sir Wilfred Grenfell Hotel housed visitors from out of town. The town had a community centre containing a movie theatre, bowling alley, library, gymnasium, and other facilities. There were also a school, fire and police departments, and a radio station. Wabush was incorporated as a town in 1967. As of 2000, the town had approximately two thousand residents, with the modern amenities of an airport, hospital, schools, churches, ski slope, and shopping and recreational facilities. It is connected by road to Labrador City, Happy Valley-Goose Bay, and Baie Comeau, Quebec. Also in 2000, the mining operations employed approximately 440 people.

Currently, the mill and mining operations at Wabush are owned by four companies: Stelco Inc. (37.9%), Dofasco Inc. (24.2%), Cleveland-Cliffs Inc. (22.8%), and Acme Metals Inc. (15.1%). The iron ore reserves are estimated at more than eight hundred million tonnes, and more than four hundred workers are employed at the complex.

AGUATHUNA

The quarry at Aguathuna on the Port au Port Peninsula began in 1911 when the Dominion Steel and Coal Company started an operation to obtain limestone for its steel mills in Sydney, Nova Scotia. Construction of a large wooden pier began in 1911. The limestone was blasted and hauled by horses to the pier and the first ore was shipped in early 1913.

In the late 1930s, the Aguathuna mine was improved with the introduction of steam locomotives and diesel equipment. The old wooden pier was replaced by a new structure in 1947, only to be destroyed in a storm the following year. Aguathuna's operations parallelled those of the Bell Island iron mines because both were supplying the same steel plant at Sydney. The quarry remained in operation until DOSCO closed it in 1964, likely in anticipation of the closedown of the Bell Island iron mine a couple of years later. Seventy men at Aguathuna were put out of work.

BAIE VERTE

Asbestos was discovered in the Baie Verte area in the mid 1950s but it was not until 1963 that the Johns Manville Company began open pit mining. The workforce at the time was approximately five hundred. By the 1970s, the health and occupational concerns relating to asbestos

*The **MV Wabana** loading limestone at Aguathuna (Courtesy of MUN CNS Archives)*

became apparent, and markets for the product dwindled. As a result the operation closed in 1981. Johns Manville sold its assets in the mine, and after 1982 there was only sporadic mining activity until the mine closed down for good in 1995.

MANUELS

For almost 100 years, a mine at Manuels has been producing pyrophyllite. This mineral has many properties and a range of applications including the manufacture of ceramic tile, brake linings, and cement. The mine property was acquired by Frederick Andrews in 1902, and production began in 1904. The company constructed an aerial tramway to the dock at Seal Cove, but went bankrupt in 1906.

In 1909, the mine was acquired by R. K. Bishop of St. John's, and operations started once again, only to cease in 1910. In 1938, Industrial Minerals Company of Newfoundland Limited resumed production at the site, shipping ore to the United States and the United Kingdom. The company ceased operations in 1947 and five years later, under the Undeveloped Mining Areas Act of 1952, the mining property was acquired by the Newfoundland government. The mine was reopened in 1956 by Newfoundland Minerals Limited, a subsidiary of American Olean Tile. It was later acquired by Armstrong World Industries Canada Limited which worked the mine until 1995. Trinity Resources and Energy Limited took it over in 1998, but as of 2002 the company was still seeking markets and had not reopened the mine.

Significant Producing Mines & Quarries in 2001

Atlantic Minerals Limited (limestone/dolomite)	*Lower Cove*
C-Mac Construction Limited(limestone)	*Cormack quarry*
Dimension Stone (dimension stone)	*various*
Epoch Rock Inc. (granite)	*Plant at Argentia*
Galen Gypsum Mines Ltd. (gypsum/sand & gravel)	*Shallop Cove, St. George's Bay*
Hi-Point Industries (1991) Ltd. (peat)	*Bishop's Falls*
Hurley Slateworks Company (slate)	*Nut Cove, Trinity Bay*
International Granite Corporation (dimension stone)	*Jumpers Brook, Mt.Peyton area*
Iron Ore Company of Canada (Iron ore)	*Labrador City*
Iron Ore Company of Canada (limestone/dolomite)	*Leila Wynne*
Lafarge Gypsum Canada Inc. (gypsum)	*Fischell's Brook, near Heatherton St. George's Bay*
Richmont Mines Inc. (gold)	*Nugget Pond, Baie Verte Peninsula.*
Shabogamo Mining & Exploration Limited (silica)	*Roy's Knob, near Labrador City*
Torngait Ujaganniavingit Corp. (anorthosite)	*Ten Mile Bay, near Nain*
Wabush Mines (iron ore)	*Wabush*

HOPE BROOK

British Petroleum Canada began developing a gold mine in the Hope Brook area on the southwestern part of the island in 1988, but stopped in 1991 for financial and environmental reasons. Royal Oak Mines, however, obtained the rights and opened a mine in mid-1992. The mine site, which was only accessible from Couteau Bay on the province's southwest coast, required the construction of five kilometres of road as well as living accommodations, a mining and processing plant, telecommunications systems, and all the other infrastructure required for an isolated mine. The gold extraction process used a heap leaching method, employing a leach pad of approximately 150 by 300 metres in size. Approximately 45,000 square metres of high density polyethylene liner were used for the pad which could accommodate up to 40,000 tonnes of ore. Sodium cyanide solution was used as the leaching agent, and the gold was recovered with a purity of about ninety per cent. The mine closed down in 1997 and Royal Oak went into receivership in 1999.

NUGGET POND

Newfoundland's only producing gold mine is operated by Richmont Mines Inc. at Nugget Pond on the Baie Verte Peninsula. It began production in 1997. In 2000, the mine employed eighty-three people and shipped 47,900 ounces of gold.

Dozens of mines have operated in Newfoundland and Labrador over the years. Most of these have been small, or were in production for only a short period of time. Newfoundland and Labrador's significant mining operations from days both past and present have been discussed in this chapter; however, the stories behind the many smaller mines beyond the scope of this volume would also make interesting material for further research.

Mining remains a significant component of Newfoundland and Labrador's economy. In 2002, the value of mineral shipments from Newfoundland is estimated to be $791 million, with iron ore representing about 92 per cent of that total. The mining industry currently employs approximately 2500 people.

Gross Value of Mineral Shipments by Commodity

($000 Canadian)

Metals	1994	1995	1996	1997	1998	1999	2000	2001e	2002f
Antimony	0	0	0	312	704	207	0	0	0
Copper	1,772	5,015	16,281	2,097	0	0	0	0	0
Gold	51,953	45,516	47,827	42,121	19,388	16,633	19,678	17,960	21,000
Iron Ore	743,137	795,839	799,331	19,409	1,026,517	760,482	902,134	754,950	728,010
Magnetite	0	2,500	10,500	3,500	0	0	0	0	0
Silver	25	38	325	107	159	138	137	80	80
TOTAL METALS	796,887	848,908	874,264	967,546	1,046,768	777,460	921,949	772,990	749,090

Non-Metals	1994	1995	1996	1997	1998	1999	2000	2001e	2002f
Asbestos	2,045	983	388	0	0	0	0	0	0
Barite	0	0	0	0	530	36	0	0	270
Brick	537	381	228	358	300	300	0	0	0
Cement	7,237	7,163	8,500	8,617	9,218	10,700	6,915	0	0
Dolomite*	3,402	2,252	3,533	8,193	7,484	7,179	9,498	6,070	5,91-
Gypsum	500	0	273	207	264	948	558	460	450
Limestone*	3,216	1,775	3,338	4,139	5,301	5,369	6,925	3,900	3,820
Peat	1,046	1,092	1,035	873	964	1,172	1,850	1,550	1,550
Pyrophyllite	1,340	1,103	4	0	0	45	5	0	1,500
Sand & Gravel	16,200	12,027	11,269	11,719	17,764	9,904	10,528	10,500	8,750
Silica	0	0	0	0	0	0	2,640	2,180	2,640
Stone (aggregate)	1,006	1,300	3,654	3,064	3,250	4,931	9,683	9,500	7,900
Stone (dimension)	4,909	4,485	4,845	5,422	3,952	3,057	4,704	7,270	9,380
TOTAL NON-METALS	41,438	32,561	37,067	42,592	49,027	43,641	53,306	41,430	42,170
TOTAL SHIPMENTS	838,325	881,469	911,331	,010,138	1,095,795	821,101	975,255	814,420	791,260

* stone (aggregate) does not include dolomite or limestone aggregate.
  Dolomite and limestone include aggregate and non-aggregate end uses.

e: estimate   f: forecast

Sources: 1993 to 2000 - Newfoundland and Labrador Department of Mines and Energy, and Natural Resources Canada;
2001e and 2002f - Newfoundland and Labrador Department of Mines and Energy (April 18, 2002).

A typical open pit mining scene (Courtesy of IOC)

Anglo-Newfoundland Development Company paper mill at Grand Falls circa 1930 (Courtesy of the Grand Falls-Windsor Heritage Society)

Chapter Six

Pulp and Paper - Serving the Newspapers of the World

The main ingredients for the production of pulp and paper are softwood, vast quantities of water, abundant hydro-electricity capability, and access to transportation. Newfoundland in the late 1800s and early 1900s was forested in softwood trees such as white pine, black spruce and balsam fir; its rivers and lakes were numerous and large enough to serve as water highways to float logs to the mills; its rivers supplied massive amounts of water which could be harnessed for hydro-electric generation; and its long coastline afforded numerous harbours which could be used as ports. Newfoundland was therefore an ideal location for pulp and paper operations. The other requirement for a project the size of a paper mill is capital, of which much would be needed for the production equipment and infrastructure facilities. The large Newfoundland mills of the early 1900s required considerable infusions of capital which had to be raised in Britain.

Construction of the penstocks for the Grand Falls power plant circa 1909 (Courtesy of the Grand Falls-Windsor Heritage Society)

NEWFOUNDLAND'S FIRST PULP MILL

The first pulp mill in Newfoundland was constructed in 1897 at Black River, Placentia Bay by the Newfoundland Chemical Wood Pulp Company Limited. This company was a subsidiary of Harvey and Company of St. John's and its main shareholders included Augustus W. Harvey, A. John Harvey, Moses Monroe, and Walter B. Grieve, all of whom were St. John's businessmen. The mill produced only about twenty tons of pulp a day, which was exported to paper plants in England. The mill however was short-lived, and because of water supply problems, ceased operation in 1903.

PAPER MILLS

GRAND FALLS

The pulp and paper mill in Grand Falls was brought about through the efforts of Alfred (1865-1922) and his brother Harold Harmsworth (1868-1940), who would later become respectively Lord Northcliffe (1905) and Lord Rothermere (1914). They had been in the newspaper business in Britain since the 1890s and were looking for secure sources of newsprint so that they would not be subjected to the vagaries and price changes of the open market and foreign supplies. Newfoundland, an English colony with vast supplies of softwood and fresh water, interested them, and in 1903 they sent a representative, Mayson M. Beeton, to look into setting up a paper mill on the island. Beeton originally considered the Bay of Islands region as a mill location with timber provided from the Grand Lake area. However, Bay of Islands was part of the so-called French shore, where France held certain rights, so he abandoned that plan. He subsequently recommended constructing a mill near Grand Falls, which would be supplied with wood from the Red Indian Lake area.

The Harmsworths accepted Beeton's recommendations, and in 1905 set up the Anglo-Newfoundland Development Company (AND) to start construction. AND had a capitalization of one million shares of five dollars each. Beeton became the company's first president. The company held discussions with the Newfoundland government regarding its construction plans and land concessions. After acrimonious public meetings and much debate in the legislature over concessions, the government of Newfoundland, under the leadership of Sir Robert Bond, approved AND's plans to develop the interior and granted the company rights to more than 2000 square miles of land, as well as water rights on the Exploits River. These rights also included all timber, minerals and aggregates in the concession area.

> **First shipment of Grands Falls newsprint**
> The first shipment of Grand Falls newsprint left St. John's harbour on January 10, 1910, on board the **SS Ulanda**.

AND, which was headed by Lord Northcliffe, began constructing the Grand Falls mill in 1907. In 1909, engineers completed construction of a large steel and concrete dam above Grand Falls, after which the town was named. The dam at about twenty-five feet in height and sixteen hundred feet in length was one of the largest engineering undertakings in the colony. Once the dam was completed, production at the mill began with the first paper rolling off the line three days before Christmas in 1909. Initially there were three paper-making machines producing thirty thousand tons of newsprint per year – two produced 256-inch wide paper, and the other, 124-inch. In 1912, two additional machines were installed, increasing the mill's annual capacity to sixty thousand tons. The seaport for the mill was Botwood, approximately twenty-two miles away, which

Lord Northcliffe (1865-1922) and *Lord Rothermere (1868-1940). The Harmsworth brothers (Alfred on the left and Harold on the right) were responsible for establishing the Grand Falls paper mill. (Courtesy of the Grand Falls-Windsor Heritage Society)*

was connected by a railway partially financed by AND. When Botwood harbour was frozen during the winter months, paper was transported by train to either Heart's Content or St. John's for shipment to England. In 1915, AND purchased two ships to transport its newsprint across the Atlantic. The ships, however, were requisitioned by the British government during World War One. During the war, production at the mill slipped to about 35,000 tons per year.

After Lord Northcliffe passed away in 1922, Lord Rothermere took over AND's operations. In 1925, a sixth paper-making machine (234 inches wide) was installed, helping boost annual production to more than one hundred thousand tons. A seventh machine, similar to the 1925 unit, was installed in 1933.

By the 1930s, most of the equipment at the Grand Falls plant was aging and newer plants elsewhere in the world were producing better quality and lower cost paper. AND therefore called in Robert A. McInnis, General Manager of one of Harmsworth's plants in Canada, to undertake a complete review of its Grand Falls operation. McInnis recommended that in order to be competitive, the plant had to be totally modernized. The company therefore implemented an improvement plan to refurbish or replace old equipment, install a new paper machine, and completely overhaul the management systems at the plant.

Surveyors or possibly engineers of the A. E. Reed Company (Courtesy of the Grand Falls-Windsor Heritage Society)

At about the time these improvements were taking place, F. J. Humphrey became the new president and recommended that the company be restructured. The old company was wound down and a new enterprise with the same name was incorporated with capitalization of $15.8 million ($7.3 million equity and $8.5 million debt). At about the same time, AND made agreements with the English *Daily Mail*, *Evening News*, and *Sunday Dispatch* newspapers to provide their newsprint needs. As of about 1937, the senior managers of AND included F. J. Humphrey, President; R. A. McInnis, Vice-President and Managing Director; V. S. Jones, Vice-President and General Manager; and J. M. Keddle, Secretary-Treasurer.

By the 1930s, Grand Falls was a well-established company town with about five thousand inhabitants. There were almost 500 company-owned houses and more than 150 privately owned residences. There were modern churches, schools and stores, and a town hall which also functioned as a movie theatre. There was also a town library containing six thousand volumes donated by Lord Rothermere and the *Daily Mail*. Other AND town sites included Botwood, Bishops Falls, Millertown, Badger, and Terra Nova, which had a combined population of around two thousand people. In the late 1940s, AND permitted residents to purchase their company-owned houses.

The Grand Falls Mill in 2001

- 220,000 ton annual capacity
- 600,000 cubic metres of wood fibre annually
- 690 mill and 600 woodlands employees
- newsprint exported to United Kingdom, Europe, South and Central America, and Asia

In 1961, AND partnered with Price Brothers and Company, a large Canadian paper-maker, to provide newsprint for the Canadian market. In 1965, Price Brothers bought the Grand Falls mill and the operation was re-named Price (Newfoundland) Pulp and Paper Limited (Price). T. Ross Moore, the former president of AND, became president of the new company. Price began numerous improvements to the Grand Falls mill. In 1967, shortly after its acquisition, the company's engineers oversaw the installation of a new state-of-the-art 650-ton-a-day paper-making machine, the most modern of its type in Canada. The four oldest units were eventually decommissioned. By the 1970s, the Grand Falls paper mill had an annual output of about three hundred thousand tons.

Price was eventually taken over by Abitibi in the mid 1970s and became Abitibi-Price. Currently the company is known as Abitibi-Consolidated Inc. and operates two paper machines at the Grand Falls mill, which produce approximately 220,000 tons of product annually.

BISHOP'S FALLS

Around the time that the Grand Falls paper mill was being set up, an English concern, A. E. Reed and Company (Reed Company) was looking into building a pulp plant in Newfoundland. The company set up in 1907, and after looking at different locations, settled on Bishop's Falls, where it completed a power plant and pulp mill in 1912. The Reed Company and AND had earlier cooperated on the construction of a railway from the Grand Falls/Bishop's Falls area to the seaport at Botwood.

The mill's output was thirty-nine thousand tons of pulp per annum, but after the outbreak of World War One, exports dropped off significantly. During the war, the company's difficulties in shipping its product to the English markets convinced it to get out of the pulp business. In 1916, it sold its Bishop's Falls operation to AND.

In order to get the pulp from Bishop's Falls to its Grand Falls mill, AND constructed an eleven and a half mile 20-inch diameter pipe between the two locations. This arrangement stayed in place until 1951 when AND consolidated pulping operations at Grand Falls and closed down the Bishop's Falls operation.

CAMPBELLTON

The relative success of the Grand Falls and Bishop's Falls plants encouraged other companies to look into the pulp and paper business. The Horwood Lumber Company of St. John's set up a small pulp mill in Campbellton, Notre Dame Bay in 1914. This plant was hardly in operation before its main dam gave way in 1915 forcing it to close down.

Bishop's Falls pulp mill during the 1930s (Courtesy of the NHS)

GLOVERTOWN

Shortly after World War One, the Terra Nova Sulphite Company, financed mainly by Norwegian investors, began constructing a pulp mill at Glovertown, Bonavista Bay. However, in 1921, financial markets lowered the value of the Norwegian kroner and the company could not afford to continue. The Norwegian company's operations were eventually acquired by AND, which later determined it would be more economical to stop construction of the Glovertown mill and proceed with an additional machine at Grand Falls instead.

CORNER BROOK

In 1915, the Reid Company set up Newfoundland Products Corporation (NPC) and obtained the water rights to the Humber River. The Reids already held the timber rights to the area through their railway agreement with the Newfoundland government. NPC was considering building a pulp mill on the river but World War One interfered with this plan. In 1921, the company entered into an arrangement with Armstrong, Whitworth and Company, which led to the creation of Newfoundland Power and Paper Company, the purpose of which was the building of a pulp and paper mill on the Humber River.

The Stadler Herter Engineering Company was hired to design the mill. The overall operation included the pulp and paper plant and town site at Corner Brook, and the hydro-power development at Deer Lake. Most paper mills throughout the world at the time were located adjacent to hydro-power facilities, but improvements in transmission line technology allowed the Corner Brook mill to be built at a port, while the electric generation facilities were located about thirty-two miles away at Deer Lake. The initial five paper machines were built by Charles Walmsley and Company, an Armstrong Whitworth subsidiary. Four of the machines produced paper 234 inches wide and the fifth was a 120 inch roll wrapping machine, producing 400 tons of product per day.

The company was anxious to complete the mill and spared no effort to get it up and running. In 1925, two years after construction commenced, the first paper rolled out of the machines.

Construction work at Deer Lake and Corner Brook employed up to seven thousand workers and cost forty-five million dollars, an incredible twenty million dollars over budget. Because of the cost overruns as well as production and marketing problems, the company fell into severe financial difficulty. In 1928, it sold its pulp and paper operations along with the Deer Lake power plant to the International Power and Paper Company (IPPC) of New York, a firm that was expanding its business through acquisitions. The Newfoundland operation became a subsidiary of IPPC as the International Power and Paper Company of Newfoundland Limited (IPPN). The year following the purchase, IPPN converted the roll

The paper mill at Corner Brook circa 1928 (Courtesy of CBPP)

wrapping machine into newsprint production and also installed two additional pulp digesters.

The company's engineers improved newsprint production from approximately 127,000 tons in 1928 to 175,000 tons in 1932. Also in 1932, IPPC's Canadian holding company, Canadian International Paper Company (CIPC), took control of IPPN. By then the Great Depression was taking its toll, and markets were dwindling, putting financial pressure on the company. Exacerbating the financial troubles was a sleet storm in 1936, which caused a broken penstock to flood the power house, putting the Deer Lake power plant out of operation. It was several months before production got back to normal.

Succumbing to heavy financial pressures, CIPC sold its Newfoundland operations in 1938 to Bowater Corporation (Bowater), a British paper company.

Sir Eric Bowater (1895-1962)
Bowater converted his family business from a paper-selling to a paper-manufacturing operation. In addition to the Corner Brook plant, which the Bowater Company purchased in 1938, the company also owned paper plants in the southeast U.S. and Liverpool, Nova Scotia. (Courtesy of CBPP)

Bowater invested heavily in the plant and quickly made the Corner Brook operations a profitable enterprise. The operation was named Bowater Newfoundland Pulp and Paper Mills Limited (changed again in 1966 to Bowater Newfoundland Limited). In 1941, two additional pulp digesters were installed as well as the Number 6 paper making machine.

The mill thrived during World War Two, and a high speed paper making machine (Number 7) was installed in 1948, making the Corner Brook mill one of the largest of its kind in the world. The new machine was built by Dominion Engineering Company of Lachine, Quebec, and produced paper at the rate of 1500 feet per minute, or 950 tons per day. Additional improvements and changes were made over the years; however the mill's largest machine (Number 3) was closed down and many employees were laid off. In December 1984, Kruger Inc. of Montreal, Quebec, purchased Bowater's Newfoundland assets and renamed the paper-making operation the Corner Brook Pulp and Paper Company (CBPP).

By this time, advances in paper-making technology had rendered most of the Corner Brook plant obsolete. In the first few years after its acquisition, Kruger spent more than $500 million to bring the plant's technology up to date. Improvements included the modernization of the paper making machines, the introduction of a recycling pulp system, and the updating of the plant's water and air quality standards. In 1986, conversion began on the mill's wood room to change it from wet to dry, and the following year the last boom of pulpwood was towed to the mill.

In 2002, CBPP manages over two million hectares of land, of which less than one million hectares is harvestable forest. The company employs 750 people at its Corner Brook mill and Deer Lake power station. An additional eight hundred to a thousand people are employed in seasonal forest operations. CBPP operates four machines: Numbers 1, 2 and 4 making 234 inch wide paper, and Number 7 making 272 inch paper. Collectively, they produce twelve hundred tonnes of paper per day.

In 1999, CBPP's Number 7 paper making machine in Corner Brook was modernized to increase its output to approximately 1150 metres per minute. (Courtesy of CBPP)

A train hauling logs crossing the dam at Deer Lake (Courtesy of CBPP)

Abitibi-Consolidated paper mill at Stephenville (Courtesy of Abitibi-Consolidated)

STEPHENVILLE

In the 1960s, Premier Joseph Smallwood encouraged the construction of a kraft liner board mill in the Stephenville area. It was thought that the Stephenville plant could be supplied by wood from Labrador and the resulting employment would alleviate the economic distress caused by the closing of the American Air Force bases at Goose Bay and Stephenville. Smallwood's ally in this venture was John C. Doyle, an American capitalist, who earlier was instrumental in starting Wabush Mines in Labrador. Construction of the Stephenville mill began in 1969, and the one thousand ton per day liner board mill was completed in 1972. From the start, the new mill experienced problems because of the difficulty in securing an economical supply of wood from Labrador.

In the fall of 1971, the Smallwood government was defeated in a general election, and a new Progressive Conservative government was installed under Premier Frank Moores. The Moores government took over the Stephenville plant and operated it for several years, but it was obvious that the mill would not become economically viable. Labrador wood shipments were discontinued in 1976 and the following year the government closed the mill down and put it up for sale. In 1978, Abitibi-Price purchased the entire operation for $45 million, a fraction of its original price. During the next two years, the company spent approximately $80 million to convert the mill into a 550 ton per day thermo-mechanical pulping newsprint mill. This huge undertaking was engineered by E & B Cowan Ltd. Comstock International was the general contractor that performed the conversion. In 1981, the Stephenville mill began producing its first paper, and as of 2002, produces about 140,000 tons of product annually.

Newsprint Shipments from Newfoundland and Labrador

Year	Tonnes	($ million)
1990	630,408	409
1991	635,989	391
1992	666,587	349
1993	703,881	419
1994	736,621	471
1995	734,660	674
1996	713,655	628
1997	740,874	568
1998	569,805	504
1999	722,185	546
2000	807,800	677
2001	745,842	676

Source: Newfoundland Forest Service, May 2002

Fisher's sawmill, near Corner Brook, circa 1922 (Courtesy of CBPP)

Edmund B. Alexander
This 669 foot long American troopship arrived in St. John's on 29 January, 1941, carrying troops and workers for the base at Pepperrell. (Courtesy of the A. C. Hunter Library, Newfoundland Collection)

Chapter Seven

Construction

Construction may be subdivided into two categories – buildings and engineering works. Buildings, of course, are the physical structures which house people, materials and other entities. Engineering works include items such as dams, power lines, bridges, highways, and refineries. Most of the province's major engineering works are discussed elsewhere in this volume, so this chapter will focus on buildings and other construction activity, including the World War Two military bases.

BUILDINGS

Early Newfoundland buildings were traditionally of wood construction and most have been destroyed through age or fire. There are still, however, many of historic significance scattered around the province, including Mallard Cottage in Quidi Vidi, the Moravian buildings in Battle Harbour, and the Ryan premises in Bonavista. Two wooden structures noted for their age and size are St. Thomas' Church in St. John's, which was completed in 1840, and the Methodist (Memorial United) Church in Bonavista, which was completed in 1923. The latter was one of the largest wooden churches in Canada at the time of its construction.

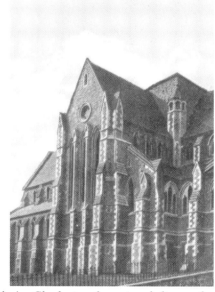

Some of the older buildings in St. John's. Clockwise from top left are the Athenaeum (burnt in the 1892 fire), the Colonial Building, the Anglican Cathedral, and Government House. (Courtesy of the City of St. John's Archives and the A. C. Hunter Library, Newfoundland Collection)

Several contemporary St. John's buildings. Clockwise from top left, the St. John's City Hall, the Mile One stadium, the Delta Hotel, and the Health Sciences Complex at Memorial University

The mid-nineteenth century saw the introduction of stone construction. Notable stone structures in St. John's are Government House (completed 1831), the Colonial Building (1850), the Roman Catholic Basilica of St. John the Baptist (1855), George Street Church (1873), St. Patrick's Church (1881), the Anglican Cathedral of St. John the Baptist (1885, with nave completed in 1850), Cabot Tower (1900), the Railway Station (1903), and the Courthouse (1904).

In the nineteenth century, St. John's was ravaged on several occasions by catastrophic fires, the most extensive in 1892. Many of St. John's buildings were razed in this conflagration, including the Church of England Cathedral. One result was the more frequent use of masonry construction in the downtown area, but many sub-standard buildings were also built to house the displaced residents before the winter set in.

In the first four decades of the twentieth century, there was little major building construction activity in the province, with the Newfoundland Hotel (1926) in St. John's, the Gambo Hotel (early 1900s) in Gambo, the Glynmill Inn (1924) in Corner Brook, and Grenfell Hospital (1926) in St. Anthony some of the notable projects. Construction activity did not pick up again until the United States and Canadian governments began building military bases in the early 1940s. These bases, including airfields, docks and buildings, will be discussed later. It was not until Confederation with Canada that other large scale construction took place in the province. The Confederation Building and Sir Humphrey Gilbert Buildings in St. John's were built in the 1950s. Other more recent major edifices include the Memorial University complex in St. John's, the construction of which began in the 1960s; the Health Sciences complex, the first stage of which was completed in 1978; numerous downtown St. John's hotels and office buildings; and a myriad of office buildings, hos-

St. John's City Engineers

Charles J. Harvey	1888-1890
F. H. Balfour (Acting)	1890-1891
R. M. Pratt	1891-1892
A. C. Waghorne (Acting)	1892-1893
T. G. Chapman	1893-1894
R. M. Pratt	1894-1896
John Ryan	1896-1920
William P. Ryan	1920-1944
Grant R. Jack	1944-1949
Ronald F. Martin P. Eng.	1949-1959
W. Duncan Sharpe P. Eng.	1959-1963
Eric W. Mercer P. Eng.	1963-1979
James Finn P. Eng.	1979-1993
Art Cheeseman* P. Eng.	1993-

*The City Engineer position was changed to Director of Engineering and Planning in 1993.

pitals, wharves and schools constructed in other regions of the province. The most recent major building projects are the St. John's Mile One Stadium and the St. John's Convention Centre, which were completed in early 2001.

As building proceeded over the centuries, the engineering associated with these construction projects changed with new technology. With early wooden buildings, the only engineering involved was civil engineering, which was required to ensure that materials were adequate, spans were the correct length and load-bearing posts were strong enough to support the weight. Later, the engineering involvement expanded when running water and sewerage became available. After the arrival of electricity, there was a need for electrical engineers to design wiring systems and their safe installation. When air conditioning appeared, mechanical engineers played a similar role. As buildings became more complex, so did the engineering. Large modern twenty-first century buildings require the skills of a large number of specialities including civil, mechanical, electrical, communications, environmental and industrial engineering, not to mention the engineering management to oversee these projects.

> **Herbert C. Burchell (1855 - ?)**
> One of Newfoundland's earliest engineers was Herbert C. Burchell, who came to St. John's from Nova Scotia. He was the Newfoundland Government Engineer between 1884 and 1905, and served as the Municipal Commission Chairman between 1898 and 1902, the modern day equivalent to Mayor of the city of St. John's. Other members of the commission included John O'Dea and Thomas White. During Burchell's tenure as Municipal Commission Chairman, Water Street was paved, the street railway was put in place, and Marconi received the first transatlantic wireless signal.

PORT DEVELOPMENT

Up until the mid twentieth century, the development of Newfoundland's harbours was limited to the construction of wharves and breakwaters to meet the needs of the fishing industry. Newfoundland's harbour developments were mainly the result of initiatives begun by the mining and paper companies for ports from which to ship their products, such as those at Botwood, Corner Brook and Bell Island. World War Two also brought on port development, especially at the US Military bases. After Confederation, federal funding helped to modernize port development in the province, most particularly at Port aux Basques and Argentia.

St. John's harbour in 1949, showing the finger piers (Courtesy of Baine Johnston Corporation)

With regard to St. John's, Newfoundland's largest port, development focussed on small projects and not the overall development of the harbour. With the exception of some new wharves constructed during World War Two, the wharves in St. John's were for the most part aging wooden structures, mainly owned and operated by Water Street merchants. The structures were simply not capable of handling the modern shipping and mechanized handling equipment of the late twentieth century. Further, many of the wharves were in an unsafe condition.

In 1956, the Federal Department of Public Works retained Foundation of Canada Engineering Corporation Ltd. (FENCO) to conduct a study of the St. John's harbour facilities. FENCO made a number of recommendations for harbour improvements, and these were accepted. The company was given the go-ahead to prepare plans and oversee the overall renovations.

Work on the improvements began in 1959. One of the main construction contracts was the dredging of 350,000 cubic yards of material from the western end of the harbour where a new finger pier, docking facilities

St. John's harbour after the improvements (Courtesy of the Port of St. John's Authority)

and large transit sheds were built. More than one million tons of rock were blasted out of the Southside Hills for use in the foundation. At the southwest end of the harbour, a docking area for small boats was built. On the north side, the wooden finger piers were demolished and a new concrete marginal wharf installed as well as a new highway along its entire length.

The harbour improvements cost almost $20 million and employed up to 1500 workers at the peak of construction. The modifications were completed in 1964.

In the 1990s, other notable St. John's harbour improvements included the dredging of the narrows to allow larger ships to enter port, and the construction of the Prosser's Rock small boat basin near the harbour's entrance.

MILITARY BASES

One of the most active periods of engineering work in the province occurred in the 1940s and 1950s. During that time, Newfoundland was a temporary home to thousands of American and Canadian military personnel. They came not only for the construction of the large Air Force, Naval, and Army bases at Argentia, Stephenville, Pepperrell, Gander, and Goose Bay, but remained for many years after the bases went into full operational service. During this period, the military also built numerous remote communications and artillery sites throughout the province.

On September 3, 1939, after the invasion of Poland, Britain and France declared war on Germany. In the spring of the following year, the British Expeditionary Force, along with Belgian and French troops, were trapped at Dunkirk in Northern France. Despite heavy bombardment by the German Army and Luftwaffe, one of the most spectacular wartime rescues was pulled off when hundreds of boats, from pleasure craft to Royal Navy warships, managed to rescue the troops by sea, and transport them safely to England.

The war was not proceeding well for Britain, and Prime Minister Winston Churchill was anxious for the United States to provide assistance. President Roosevelt's hands were tied because of America's Neutrality Act which forbade the country from assisting in a war where it was not involved. The British and Americans came up with a plan to circumvent this law and allow the United States to indirectly provide assistance to Britain in the way of ships and equipment. Thus was born the Lend Lease arrangement whereby Britain would lease territory to the United States in Newfoundland, Bermuda, and the West Indies for military bases, and in return receive destroyers from the Americans.

In September 1940, American armed forces began arriving in St. John's and by year-end were also setting up in Argentia. By the end of the war, more than one hundred thousand American army, navy, marine and air force personnel would have been stationed in Newfoundland, providing it with an economic and construction boom not witnessed before. During and shortly after the war, more than 750,000 American personnel, 45,000 aircraft, and 10,000 ships passed through the Goose Bay, Gander, Harmon Field, Argentia and Torbay bases.

ARGENTIA NAVAL BASE

On October 13 1940, the first US Military contingent arrived at Argentia, Placentia Bay to survey the site for a naval base. The location was ideal because the town's deep harbour was the largest in Newfoundland that was both close to Europe and ice-free year round. Prior to construction of the base, Argentia was a fishing village of about five hundred inhabitants with a connection to the main Newfoundland railway system at Placentia, just a few miles east of the town. It was also served by coastal steamers which regularly stopped at the port. On January 18, 1941, the quiet town was disrupted when US Naval ships arrived with fifteen hundred construction workers. This initial contingent was followed a week later by 120 Marines to provide security and police functions at the site. The US personnel lived on the *USS Richard Peck*, which was used as a dormitory until 1943.

Construction of the Argentia base resulted in the displacement of the local residents. To make room for the facility, the Newfoundland government used its wartime powers and expropriated the land in the area. The US Military compensated the inhabitants affected and located them to nearby towns. Not only citizens were affected, as the graves from the local cemeteries were transferred as well.

The initial construction contract for the base, valued at almost $9.5 million, was awarded to Fuller Construction Company on January 28, 1941. Part of the work was the removal of 8.5 million cubic feet of

An aerial view of the Argentia Base with the docking facilities at the top of the peninsula and the runways in the centre. The Fort McAndrew army base is off the peninsula to the rear. (Courtesy of MUN Maritime History Archive)

peat and gravel. By the end of 1944, additional contracts had pushed the total construction cost to more than $53 million. Up to about fifteen hundred Newfoundland construction workers were hired to help build the base. At the time, Argentia was the most expensive military establishment built by the United States outside the country.

The Argentia base, which comprised 3392 acres, consisted of a US Naval Station with seaport and airfield facilities, as well as Fort McAndrew, an Army facility. Construction included the building of aircraft runways, wharves, and accommodation facilities. The work was

US Naval Dock at Argentia during World War Two (Courtesy of MUN Maritime History Archive)

completed in 1941 and Argentia was fully operational by the time the United States entered the war after Pearl Harbour was bombed on December 7 that year. During the war, Argentia was home to five aircraft carriers and more than fifty destroyers and support vessels. The Naval Air Station had three runways which supported transatlantic ferry operations as well as reconnaissance and fighter aircraft. The base was also home to part of Britain's Royal Navy fleet.

After the war, expansion of the base continued and included what was at the time Newfoundland's highest building, the ten storey high Bachelor Officer's Quarters. This building was begun in 1955 and was used up until the base was closed, only to be demolished by implosion in 1999. After the war, Argentia continued to be used for a variety of military purposes, but by the early 1970s it was beginning to outlive its strategic usefulness. The Air Station closed down around 1973, followed by the Naval Station in 1994.

FORT PEPPERRELL

On October 13, 1940, a little over a month after the Lend-Lease Agreement was ratified, the American naval ship *USS Bowditch* arrived in St. John's with a contingent of the US Army Corps of Engineers to begin preparatory work on a new base. The main site was on 191 acres on the north side of Quidi Vidi Lake. At the time, the area was occupied mainly by small farms and the US Army made financial arrangements with the inhabitants to relocate them to other areas. All told, about 228 acres of land in the St. John's area were passed over to the US Army, including the land at Quidi Vidi Lake, as well as smaller parcels on Signal Hill, the White Hills, and near Radio Tower Road.

The first contingent of construction workers arrived in St. John's on January 29 1941 aboard the *Edmund B. Alexander*. The 669 foot long ship was the largest to enter the harbour up to that time, and carried 977 US Army troops, as wells as two thousand tons of construction material and military hardware. A temporary camp was set up in the north-central area of St. John's between Carpasian Road, Pine Bud Avenue, and Rennie's River to house the troops until the Pepperrell base near Quidi Vidi Lake was constructed. This fifteen-acre area became known as Camp Alexander.

The first military site built in the St. John's area by the Americans was on Signal Hill, near the site of the present day Signal Hill interpretation centre. This establishment included barracks, mess hall, recreation

An aerial view of Fort Pepperrell in St. John's (Courtesy of MUN Maritime History Archive)

facilities and artillery placements. It was completed by May 1941 and about one thousand troops were eventually stationed there. Earlier, in February, the work on the Fort Pepperrell Base began and was largely completed by the end of the year. The overall contractor for the project was Newfoundland Base Contractors which managed the construction along with the US Army Corp of Engineers. Approximately fifty-five hundred civilian workers were employed during its peak construction. By mid December, all troops living at Camp Alexander had moved into the permanent quarters at Fort Pepperrell.

    In October 1941, the US Army began construction of a naval dock and storage sheds at the east end of St. John's harbour. The six hundred foot long dock could accommodate ships with a draft of up to thirty-one feet and was primarily used to handle shipping associated with supplying the Pepperrell Base. The facility first saw service in 1942, but was not fully completed until 1943.

    Meanwhile, the Royal Canadian Air Force was setting up an airbase at Torbay airport which became operational in December 1941. This facility was also used by the US Army Air Corps (USAAC) and the Royal Air Force. In early 1947, the RCAF turned over operation of Torbay to

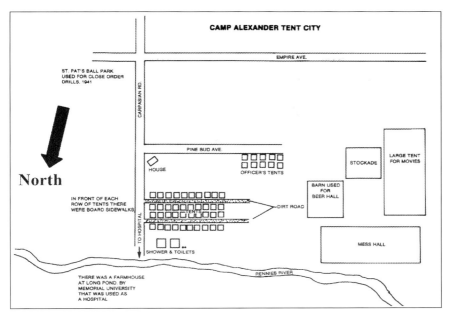

Camp Alexander
The temporary home for the builders of Fort Pepperrell, now a prime real estate area in St. John's. Please note the unnusual orientation of the map. (Courtesy of MUN Maritime History Archive

the USAAC; however, six years later, it moved back and shared facilities with the American forces. The USAF, which absorbed the US Army Air Corps in October 1947, remained at Torbay until 1958.

On January 1, 1946, Fort Pepperrell became Pepperrell Air Force Base. This facility closed down in 1960 and its 208 buildings were passed over to the Canadian government. The buildings included a hospital, movie theatre, radio station, telephone exchange, and a number of apartment buildings.

HARMON AIR FORCE BASE

The Stephenville Air Base, an Army Air Force facility, was the third US Military base constructed in Newfoundland. Its location on more than eighty-one hundred acres of land near Stephenville, St. George's Bay, was relatively free of fog and an ideal location for refuelling military aircraft flying to and from Europe, Greenland and Iceland. In January 1941, the US Army Corps of Engineers arrived, followed a month later by 150 men of the US Signal Corps, to set up communications links with Fort Pepperrell.

Stephenville at the time was a typical small Newfoundland town relying on fishing, farming and lumbering. Its population was around five hundred. As with the other bases, the US government provided compensation to the residents who had to be relocated. Newfoundland Base Contractors again headed the construction and after the first contract was awarded in February 1941, work commenced on buildings, hangars, and runways. During construction, the tradesmen and Army personnel lived in a tent city until the permanent quarters were completed.

In June 1941, the Stephenville Air Base was named Harmon Field, in honour of Captain Earnest Harmon, an American pilot who was killed in an air crash. The base was originally planned to be a temporary facility and the construction was of a standard reflecting this. The runways were only partially completed in 1942 and were used initially for emergency operations. They did not open for heavy air traffic until 1943. During the peak construction period, the civilian population in the town of Stephenville exploded to more than seven thousand inhabitants.

During the remainder of World War Two, the base played a pivotal role as a servicing and refuelling point for aircraft ferrying personnel as well as for bomber and fighter aircraft involved in the European conflict. When the war ended, it served as a major stopover point for the thousands of American troops returning home.

In 1947, the US Military decided to upgrade the Harmon facility to make it a permanent Air Force Base. Modern buildings were constructed to replace the wartime structures. On July 1, 1948, the base was renamed Ernest Harmon Air Force Base. In 1953, further improvements were made to lengthen and improve the runways. New buildings, including a one hundred bed hospital, were also constructed, as were new aircraft maintenance facilities to serve the larger and newer aircraft used during this era. The upgrade was part of the United States commitment to the North American Aerospace Defence Command (NORAD), to guard the northeastern North American flank from a surprise attack by Soviet bombers and missiles.

At the end of 1966, the USAF closed down the facility, which became the last US base in Newfoundland to deactivate. The USAF turned over its remaining assets to the Canadian government, which set up the Harmon Corporation to oversee their use and disposition. The air-

View of Harmon Air Force Base (Courtesy of MUN Maritime History Archive)

port continued as a commercial airport: Air Canada, which started commercial flights in and out of Stephenville in 1949, operated a commercial service there until 1989. Eastern Provincial Airlines and its successor airlines also provided service to Stephenville until the late 1990s. In 2002, the airport is used for domestic flights, and as a refuelling stop for corporate jets and the occasional transatlantic flight.

GANDER

In 1936, the Newfoundland Commission of Government began building an airport near present day Gander. The initial construction included a hangar, several support buildings, and a single unpaved runway suitable for small and medium size aircraft. On January 11, 1938, Captain Douglas Fraser, who was the first pilot hired by the Newfoundland government, made the first official landing at the airfield.

At the beginning of World War Two, the Royal Air Force was in dire need of aircraft and had to import them from Canada. In actual fact, the bombers were manufactured in the United States but because of the US Neutrality Act, could not be flown across the US-Canada border and

A group of more than one hundred B-17s waiting at Gander for delivery to Britain during World War Two (Courtesy of MUN Maritime History Archive)

had to be hauled across by horses or tractors. Transporting these new aircraft to Britain by sea would have been risky because of the German U-Boat menace. They therefore had to be flown, and refuelling stops would be required. Gander was an obvious choice for one of these stops. Consequently, in 1940, the Royal Canadian Air Force took over operation of the airport to handle these flights. The RCAF built a new control tower, hangar, and runways, and lengthened the existing runway. One of the runways was also paved. In November 1940, a fleet of seven Hudson bombers made the first flight from Gander to Ireland, the first of many such flights the Atlantic Ferry Command would undertake throughout the war. After entering the conflict, the United States also used Gander and later, Goose Bay in Labrador, as refuelling stations for its overseas military aircraft. To better serve their aircraft, the US Military entered into an agreement with the RCAF and the Newfoundland Commission of Government in late 1941, to locate its own facilities and personnel at Gander. During the war more than ten thousand aircraft were "ferried" via Gander and Goose Bay to Europe.

The Gander air base involved the participation of numerous civilians in support roles. Civilian employment during the war was approximately fifteen hundred, all of whom required shelter and support infrastructure. The town of Gander therefore had its birth. Its prominence as an airport town has continued to the present.

When World War Two ended, Gander also played an important role in servicing aircraft carrying material and personnel back to the United States and Canada from Britain and Europe. By 1948, after thousands of flights, the US Military pulled out of Gander, followed shortly thereafter by the RCAF, which nevertheless kept a small contingent to service the occasional military flight. Since the end of the war, the airport expanded and became a major commercial international airport. Gander was a major refuelling stop for transatlantic passenger lights until the 1990s, but now only services domestic airlines as well as occasional transatlantic flights.

GOOSE BAY

By 1941, the Gander airport was overburdened by the number of aircraft using the facility as a refuelling stop, so the USAAC looked at alternative sites. Under the direction of Captain Elliott Roosevelt, the son of the US president, other locations were examined and Goose Bay, Labrador was selected. Goose Bay was also the site favoured by the RCAF which sent engineers there to begin planning the new air base.

McNamara Construction Company of Montreal was awarded the main contract, and material and supplies began to arrive in mid September 1941. By the first week of November 1941, construction of the first runway was advanced enough to allow aircraft to land. At year end, there were more than three thousand RCAF personnel stationed in Goose Bay. By early summer of the following year, two three thousand foot concrete runways had been completed, as well as temporary buildings, warehouses and gasoline storage tanks.

In 1942, the United States obtained approval for a base at Goose Bay, and by June, several dozen American personnel were at the site. By the end of 1942, there were more than three thousand construction workers on the base as well as approximately five thousand Canadian, American, and British personnel.

The US began ferrying aircraft to Britain in June and by the end of 1942 had handled 662 planes through Goose Bay, not to mention those handled by the RCAF. In 1944, 5262 US aircraft were ferried to Britain via Goose Bay, and over the duration of the war approximately twenty-five thousand Allied aircraft were handled at the airport.

Over the next thirty years, numerous improvements were made to the Goose Bay base. The construction of new hangars, fire stations, and warehouses, as well as runway improvements, continued on both sides of the facility. By 1967, however, the strategic importance of Goose Bay was diminishing. The RCAF turned the operation of the airfield over to the Canadian Department of Transport (DOT) and the few remaining RCAF personnel were relocated to the American side. In July 1973, the Americans also turned their buildings and facilities over to DOT and officially closed down their base on October 1, 1976. It is estimated that the net worth of the Goose Bay base at the time was about $250 million.

In the 1980s, the RCAF reactivated the base at Goose Bay as a NATO training facility for the testing of low level flying. Every year, several thousand low level flights are conducted by NATO air forces, including those of the United Kingdom, the Netherlands, Germany, Italy, and Canada. These flights are carried out at elevations as low as one hundred feet. A 130,000 square kilometre test area in central Labrador is also used for aircrews to practise dropping non-explosive bombs. In the meantime, Goose Bay continues as a commercial airport for domestic flights as well as occasional international refuellings.

COMMUNICATIONS AND ARTILLERY SITES

During its presence in Newfoundland, the American military built many small structures throughout the province. With lookout towers, artillery sites, repeater sites, radar sites, and Air Control and Warning sites, most communities in Newfoundland were not far removed from a military installation.

Overall, there were about eleven artillery sites, thirteen look-out towers, five transmitter sites, five direction finding stations, ten repeater stations, five radar sites, and seventeen Air Control and Warning radar gap filler stations. There were two significant communications links, both built by the USAF – the Pepperrell-Argentia-Harmon Field pressurized communications cable, and the Pepperrell-Stephenville and Pepperrell-Frobisher troposcatter systems.

The first of these was a 504 mile long pressurized cable which provided up to 44 circuits. The cable consisted of a number of copper conductors enclosed in a lead sheath which was pressurized at various

Troposcatter Antenna at Cartwright, Labrador
This site was part of the United States Ballistic Missile Early Warning System, which was built in the 1950s. (Courtesy of Newtel)

points along the route with nitrogen gas to keep out moisture. The prime contractor for the project was Bell Canada, which began construction in the fall of 1942 and finished in the spring of 1943, installing up to ten miles of cable per day. In addition to the terminals at Fort Pepperrell, Argentia, and Harmon Field, repeater stations were located at Whitbourne, Shoal Harbour, Gander, Grand Falls, Millertown Junction, Howley, and Corner Brook.

The second major project, the Pole Vault troposcatter system, interconnected American radar sites in Newfoundland and Labrador and Baffin Island with military headquarters in the United States. Sites were constructed at Frobisher, Resolution Island, Saglek, Hopedale, Goose Bay, Cartwright, St. Anthony, Gander, Harmon Field, and Fort Pepperrell. The system used huge parabolic antennas between sixty and one hundred twenty feet in length, transmitting at a power of ten thousand watts, and providing up to thirty-nine circuits. The system was completed in 1955. After the USAF decommissioned it in the 1960s, several sites were taken over by Avalon telephone, which used part of the system for telephone circuits until the last sites were closed down in the 1970s.

Chapter Eight

Marine - Fishery and Shipbuilding

From the time John Cabot discovered Newfoundland, the Grand Banks area off the province's south east coast has been recognized as one of the world's largest fishing grounds, providing fish for Europe, Canada, the United States and parts of the world as far away as Japan. Fleets from many countries have harvested the Grand Banks - among them fleets from Newfoundland, Canada, Great Britain, the United States, Russia, France, Poland, Japan, Norway, Portugal and Spain.

Largely because it was discouraged by the English government, there was little in the way of settlement on the island of Newfoundland in the sixteenth and seventeenth centuries. The British preferred the "migratory fishery," in which ships departed England for Newfoundland in the spring and returned with their cargoes of fish in early fall. One reason for this was the country's

> **SS Blue Peter**
> In 1928, Job Brothers purchased the **Highland Laird**, a factory refrigeration ship, and renamed it the **SS Blue Peter**. For about ten years, this ship travelled throughout Newfoundland waters, purchasing and freezing salmon for the English and European market. This was the first refrigeration ship used in the Newfoundland fishery.

desire to maintain at home in Britain a mass of able-bodied seamen who could be pressed into the Royal Navy if and when the need arose; another was the fish merchants' desire to maintain their control over the fishery.

An automatic weighing machine at the FPI coldwater shrimp processing plant in Port Union (Courtesy of Fishery Products International Limited)

In the early eighteenth century, sailships in the migratory fishery would arrive in Newfoundland in the spring. Some would harvest a load of cod and head back to port. Others would drop ashore fishers who would fish from small boats, and salt-cure and sun-dry the fish on land. They would be picked up and returned to Europe in the fall with their harvest. The main markets for dried Newfoundland cod were Spain, Portugal and Italy, and although the Newfoundland fishery was prosecuted mainly by English concerns, there was only a small market in England for Newfoundland cod.

Despite the English government's discouragement of permanent settlement, a number of fishers did live in Newfoundland throughout the year. These people often lived close to good fishing areas in small coves and harbours that were not easily detected by the British Navy or meandering pirates. By the early 1800s, Newfoundland's population was around one hundred thousand. Up until then the fishery had been mainly prosecuted by firms from the other side of the Atlantic. These large fish merchants, primarily those from the west country of England, eventually decided that it was more advantageous for them to locate in Newfoundland year round, so they began to establish premises in the colony.

One of the earliest mercantile firms in Newfoundland was set up in the 1740s at Trinity, Trinity Bay, by Benjamin Lester from Poole, Dorset. Although not originally arriving in Newfoundland as a merchant, he inherited "fishing rooms" from his father-in-law, and in partnership with a brother who was managing the family business in Poole, became the wealthiest merchant in Newfoundland. Some other merchants of the period were Slade (in Fogo), Munn (in Harbour Grace), Ridley (in Harbour Grace), and Ryan (in Bonavista).

The English merchants operated a kind of barter system with the local fishers. In return for fish, the fishers were given credits which they could only use to purchase food, gear and supplies from the same merchant. There was little cash exchanged, and usually the fishers were lucky if they could eke out a subsistence-level existence. The merchants controlled the fisheries in their respective locations in the colony, and thousands of fishermen depended on them for credit, as there was no other way for them to be paid or to get their product to the markets in Europe.

FISH PROCESSING

Because Newfoundland's fish were so far from the European markets, it was impossible to deliver fresh cod before it spoiled. The only people who had the luxury of fresh Newfoundland cod were the planters and the fishers aboard the ships. In the twentieth century, freezing technology helped eliminate the problem, and when modern air transport became available, fresh fish product (especially lobster) could be easily, although expensively, shipped to European and North America markets.

SALTING

Up until the early twentieth century, salting was the traditional method of processing and preserving Newfoundland cod. Fishers who did not come ashore took part in the so-called "green fishery," in which the fish were gutted and cleaned aboard ship and stored in salt in a ratio of about sixty pounds of salt to one hundred pounds of cod. When the holds were full, or the season was ending, the ships would head towards their market destinations. The fishers who resided on land during the fishery split and cleaned the cod and salted them. The fish were air-and-sun-dried on outdoor flakes. Natural drying was the norm until 1947, when George Dixon built Newfoundland's first artificial drying plant in the Burin Peninsula community of Fortune.

With the exception of various fish species that were pickled or smoked, salt cod remained the main fish product until freezing came on the scene. Mechanical drying quickly became the standard for exported salt fish. With help from the Canadian Saltfish Corporation, salt fish continued to be produced up until the early 1990s.

FREEZING

The most significant advance in fish processing was the introduction of freezing technology during World War One. Freezing added a new element to the fishing industry because to a much larger extent than salting, it required specialized plants to handle, process, package, freeze, and ship the product. Engineers had to design processes to work efficiently and economically, a challenging task since different species required specialized approaches.

In 1916, Harvey and Company of St. John's built a bait freezing plant at Rose Blanche on the southwest coast of the island. Freezing plants were also constructed at St. John's and Bay Bulls the following year. In the 1920s, Job Brothers installed freezing plants at Bonavista, St. Anthony and Englee. The first quick-freezing plants appeared in Newfoundland during World War Two. This type of freezing used a very low temper-

> **Quick Freezing**
> This process was invented by Clarence Birdseye (1886-1956), an American inventor, who had spent some time trapping and fur farming in Labrador during the 1910s. He noticed that fish and other food products which quickly froze during a Labrador winter tasted better than those frozen by mechanical means. In the 1920s, in the United States, he perfected a quick freezing system which is the basis for today's modern freezers.

ature, as well as forced air, to quickly freeze the fish, making it taste more like a fresh product. World War Two dramatically increased the production of frozen fish, with almost all the product going to England. By 1943, there were fourteen fish refrigeration plants in Newfoundland, increasing to eighteen in 1945. Frozen fish was a popular commodity for the next few decades especially for export to the United States. Large companies such as Bonavista Cold Storage, Fishery Products Limited (later Fishery Products International Limited), Lake Group Fisheries, and National Sea Products led the way in the frozen fish industry, but there were also smaller companies such as Bay Bulls Sea Products, Newfoundland Quick Freeze, Ocean Harvesters, T. J. Hardy, and Quinlan Brothers that made Newfoundland a major fish distributor. In 1997, ninety-nine plants had

The FPI crab processing plant in Triton, Notre Dame Bay (Courtesy of Fishery Products International Limited)

groundfish freezing capability. With the industry now in the process of restructuring, that number had decreased significantly by 2002.

DEMISE OF THE COD FISHERY

The frozen fish industry quickly expanded in the early 1950s, creating an oversupply in the United States market which adversely affected prices. Market conditions improved in the 1960s but during that period, foreign fishing also increased significantly. In the 1970s, ships of all nations were using the most efficient means that technology permitted to harvest the maximum amount of fish from the Grand Banks. And despite controls, it was difficult to police or prevent overfishing of the resource, especially on the part of foreign trawlers. In 1977, Canada implemented a 200-mile economic zone around its coastline to manage, police, and if necessary, arrest and prosecute trawlers which were illegally fishing. In the same year, only 77,000 tonnes of northern cod were landed in Newfoundland, compared to the approximately 150 to 200,000 tonne range in the 1960s. Over the next fifteen years, however, catches decreased, and it became apparent that the fish stocks were being depleted.

In 1992, as a shock to all Newfoundland fishers and fish plant workers, the Newfoundland fishery came to a screeching halt, when the Honourable John Crosbie, in a presentation at the Radisson Hotel in St. John's, announced a moratorium on the Northern cod fishery. Thousands of fishers and fish plant workers were thrown out of work in the most devastating economic blow in Newfoundland's long history. As of 2002, the cod stocks had not returned, and it is not known when, or even whether, fishing for cod will ever recommence. Since the moratorium, major fish producing companies in the province, such as Barry's and Fishery Products International, have been forced to refocus their marketing strategies and develop and market other sea products, the most significant being crab and shrimp. Shellfish – crab and shrimp – has almost entirely replaced groundfish as the major fishing sector and in many cases has proven to be more profitable for the fishers and processors.

AQUACULTURE

One of the earliest recorded aquaculture projects in Newfoundland was at Dildo Island, Trinity Bay where the government built a cod hatchery in 1889. This project was overseen by Adolphe Nielsen, a Norwegian hired by the government as a consultant, and was designed to produce between two and three hundred million cod ova per year. A year or two later, a pond was built which was filled with sea water pumped from the ocean with the aid of a 5.5 metre windmill built for that purpose. In 1893, the facility produced more than 200 million cod fry, as well as more than 500 million baby lobsters, which were released into the nearby bay. For unknown reasons, the facility shut down after operating only a few years. Since then, and especially since the 1992 cod moratorium, the aquaculture industry has grown. In 2000, aquaculture was conducted at 196 sites in the province, exporting almost $14 million worth of product, including salmon ($5 million), steelhead trout ($5.5 million), and blue mussels ($2.7 million).

EARLY SHIPBUILDING

As Newfoundland is surrounded by thousands of miles of coastline, it is to be expected that shipbuilding would be a significant industry. Shipbuilding has passed through a number of phases, all the way from the building of small punt-like rowboats to the fabrication of offshore vessels for the oil industry.

Fishers began building their small boats in Newfoundland shortly after they started to settle on the island, as it would have been impractical to bring boats from Europe. The earliest recorded instance of shipbuilding in Newfoundland was at John Guy's colony at Cupids in 1611, when he reported in his log that "A boat, about twelve tons big, with a deck, is almost finished to saile and row about the headlands." Mentions of shipbuilding in the records of the late seventeenth and early eighteenth century are sketchy. It is clear however that fishermen made their own fishing boats for inshore use, and that thousands of such vessels were built.

One of the first instances of organized shipbuilding was at Trinity during the mid 1770s. Benjamin Lester, one of the early West Country merchants to locate in the town, had a number of ships built there and for that purpose brought over Charles Newhook, a master shipbuilder from Poole. The Newhook name has been synonymous with shipbuilding in Newfoundland for generations, especially at New Harbour, Trinity Bay.

Also in the late 1700s, ships were built at Heart's Content, Carbonear, Harbour Grace, Cupids and Brigus; these were mainly schooners around fifty tons. During the winter of 1809-1810 (most shipbuilding was done during the winter), three schooners as well as a brigantine were built in shipyards at Trinity, not to mention another twenty-five small boats built by local fishermen.

In the 1800s, many of the larger Newfoundland merchants owned shipyards, and built ships for their own purpose or for sale to others. Thus John Rorke of Carbonear, John Munn of Harbour Grace, and Charles Bennett of St. John's became major shipyard owners of the era. Smaller shipyards also sprang up in other locations around the colony including Notre Dame Bay and Halls Bay. Just about all the shipbuilding effort in Newfoundland in the 1800s was directed to sailing vessels, although later in the century, steam vessels were also constructed.

TWENTIETH CENTURY SHIPBUILDING

In the first quarter of the century, there was a flurry of ship building, especially in the period around World War One. Between 1917 and 1920, approximately forty schooners in the three to six hundred ton range were built. These were constructed in various locations including St. George's, Bay of Exploits, Charlottetown, Port Union, Placentia, Port Blandford, and Harbour Grace. The last schooner constructed in Newfoundland was the *Alberto Wareham*, which was built by Thomas Hodder of Marystown in 1949.

World War Two brought about another era in Newfoundland shipbuilding when the Commission of Government established a shipyard at Marystown. The yard was begun in 1939 and before being hit by a disastrous fire in early 1941, had produced four minesweepers. The following year, another government shipyard at Clarenville began production of the so-called "splinter fleet," ships which were originally designed as minesweepers, but later used for local freight and passenger service.

The second half of the twentieth century saw the introduction of "the longliner," a decked motorized fishing boat between thirty-five to sixty-five feet in length, used mainly for the cod-trap fishery. Hundreds of longliners have been built in many areas of the province. Most of these were of wood construction, but some were of steel and others were fibreglass-reinforced.

In 1966, the Newfoundland government set up a crown corporation to operate the shipbuilding and ship repair facility at Marystown. Canadian Vickers, a large Canadian shipbuilding concern, was contracted to operate the facility; however, because of a downturn in the shipbuilding industry, the company decided to pull out of the arrangement in 1971.

The government hired John Rannie, a Scottish shipyard executive, to head the crown corporation. Within a few years, the shipyard had built fifteen trawlers, including seven for National Sea products, several offshore oil vessels for the North Sea oil industry, and a patrol vessel for the Canadian government. In the 1980s, several offshore supply vessels were constructed for the Newfoundland oil industry, as well as ferries for the Newfoundland government, including the fifty-three-metre-long *Beaumont Hamel*, built in 1985, and the fifty-four-metre-long *Flanders*, built in 1990. After the cod fishery collapsed in the early 1990s, there were no new orders for trawlers and the workforce at the shipyard dropped to fewer than one hundred workers. As of 2002, fifty-six ships had been built at the shipyard – the latest, the 37.92 metre escort tugs *Placentia Hope* and *Placentia Pride*, which were delivered to Newfoundland Transshipment Limited in 1998.

Marystown Shipyard (Courtesy of Peter Kiewit Sons Ltd.)

The government spent $40 million to upgrade the facility, and in 1993, the shipyard unsuccessfully bid on a contract for work on the Hibernia platform. Within a year, however, Marystown's bids on other construction contracts for the Hibernia project were successful. In 1996, the shipyard completed construction of two offshore supply vessels, the *Maersk Nascopie* and the *Maersk Norseman*, each 82.3 metres long with a 3500 ton displacement, the largest ships ever built in the province.

The Marystown shipyard has more than one thousand feet of water frontage, a 2448 by 59 foot syncrolift platform dock, equipped with twenty 180 tonne hoists, as well as a 101,000 square foot fabrication area. A marine facility was started up in the early 1980s at Cow Head, about seven kilometres north of the shipyard. Initially designed to service offshore oilrigs, it expanded in the 1990s as a fabrication facility. Cow Head can now accommodate offshore drilling rigs for servicing, and also perform fabrication work at its 820,000 square foot facility.

Because the Marystown and Cow Head facilities were losing money and could not survive without government subsidies, the government turned both over to Friede Goldman, a large international offshore company, for the sum of one dollar in late 1997. One of Friede Goldman's obligations was to guarantee a certain level of employment, which it could not meet. However, in 2001, the company filed for bankruptcy protection. In 2002, the situation was resolved when Peter Kiewit Sons Ltd.,

Cow Head fabrication facility near Marystown (Courtesy of Peter Kiewit Sons Ltd.)

a large American offshore concern, purchased Friede Goldman's assets and announced that it would continue the Marystown and Cow Head operations and pursue contracts on the White Rose project.

SHIP REPAIRS

In addition to the Marystown Shipyard, the other large ship repair facility in the province is located at St. John's, where St. John's Dockyard Limited (NEWDOCK) operates a ship repair, offshore services and industrial facility. The history of NEWDOCK's predecessor companies began in the late 1800's with the first St. John's dry dock. This facility, after many improvements, is still in operation. In addition to its 174 metre dry dock, the company also operates a 4000 ton marine elevator. NEWDOCK has a manufacturing and fabrication area of 6550 square metres with milling, welding and other fabrication equipment along with extensive sub-sea testing facilities for the offshore oil industry. The company has approximately five acres of laydown area and cranage capacity of up to eighty tons.

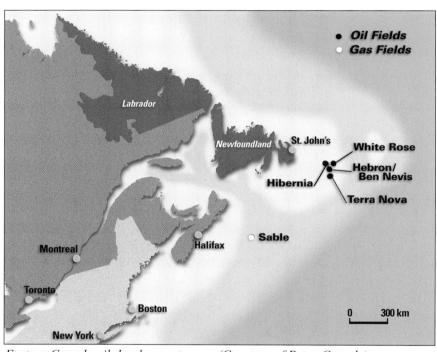
Eastern Canada oil development areas (Courtesy of Petro-Canada)

Chapter Nine

Oil and Gas - From Parsons Pond to Hibernia and Beyond

From its early days of settlement, Newfoundland and Labrador relied primarily on resources from the sea. Later, in the 1800s, its copper and iron mineral resources were exploited, and around the turn of the century, the colony's hydro electric potential began to be developed. A few years later, Newfoundland's forests provided the raw materials for the establishment of paper mills. The province's oil resources, however, have only recently seen development, which did not really begin in earnest until the late twentieth century.

Oil was traditionally derived from whales and seals, as well as cod livers; it would not be until the mid nineteenth century that Newfoundland produced its first hydrocarbon oil. In the mid twentieth century Newfoundland and Labrador saw its first modern refineries; later, offshore hydrocarbons were discovered off the coast of Labrador and on the eastern edge of the Grand Banks. The Grand Banks oil and gas resource is proven to have enormous economic potential, and has resulted in some of Newfoundland's most remarkable engineering accomplishments.

REFINERIES
HOLYROOD

One of Premier Joseph R. Smallwood's plans was to industrialize the province and make it less dependent on oil imports. In May 1960, Smallwood announced that his government had an agreement with the Golden Eagle Refining Company of Canada for the construction of a refinery at Holyrood. Part of the deal called for the government to purchase all of its petroleum needs from the refinery for a period of twenty years. The president of Golden Eagle Refining was John Shaheen, whose name would also rise in connection with another refinery at Come by Chance. The $20 million Holyrood complex opened on December 21, 1961. In addition to the fifteen thousand barrel per day refinery, the operation also included wharf and storage facilities as well as a spur line connecting with CNR's railway tracks.

The twenty-year arrangement with the government expired in 1981. After that, the company had problems finding markets for its products and in 1983 its refinery was operating at less than 50% capacity. With this level of production the operation was not economic and in June 1983 it closed down. In the following year, the plant was dismantled.

COME BY CHANCE

After the initial success of the Holyrood refinery, the Smallwood government was interested in further expanding the province's refining capacity. It again entered into discussions with John Shaheen, who had built the Holyrood refinery. Shaheen reached an agreement with the government to construct an oil refinery which would have a production capability of more than one hundred thousand barrels per day. In April 1967 Smallwood introduced legislation for the financing of the refinery which was to be located at Come by Chance, Placentia Bay, a deep ice-free harbour near transatlantic shipping routes. The government guaranteed Shaheen a $30 million debt issue and provided $5 million of bridge financing. The financial details caused a political furor in the House of Assembly and resulted in John Crosbie and Clyde Wells, two of Smallwood's senior cabinet ministers, crossing the floor to sit with the opposition. Both of these men would later take places of prominence on the political scene. After several years of wheeling, dealing, and somewhat suspect agreements with the government, $155 million of financing

North Atlantic refinery at Come by Chance (Courtesy of North Atlantic Refining)

was arranged with a British bank. The Federal government also contributed to the project by building a $20 million wharf and docking facility. At about the same time, a ten year agreement was arrived at with the British Petroleum Trading Limited for the provision of one hundred thousand barrels a day of crude oil.

The official opening of the refinery was held in October 1973, and Shaheen invited several thousand guests. Come by Chance had limited accommodation, so he hired the liner Queen Elizabeth II as a floating hotel for the opening ceremonies.

The original design of the refinery called for the production of different types of fuel, including automobile gasoline of several different octanes; jet fuel; kerosene; and other petroleum products. Sulphur, a by-product, was also to be shipped as a commercial product. Two 31,000 ton tankers were purchased to service its operations, named somewhat ironically after two political foes of the day, *Joseph R. Smallwood* and *Frank D. Moores*.

After the refinery began commercial operation, design flaws were identified leading to production problems. Some corrections were implemented, but Shaheen was by then experiencing cash flow problems, and not all work was completed. The company misjudged its crude carrier requirement and over-contracted its needs. Further, the actions of Arab oil producers in October 1973 greatly increased the cost of crude. These and other factors forced the refinery into receivership in 1976 with a debt of more than $500 million, the largest bankruptcy in Canada at the time.

The refinery lay in mothballs for a few years and it was not until April 1980 that the receivers found a buyer in Petro-Canada. The purchase price was $10 million. Petro-Canada eventually decided that reactivating Come by Chance would not be part of its business plans and decided to put the refinery up for sale. In September 1986, it sold the facility for the princely sum of one dollar to Newfoundland Energy Limited, a Bermuda based company, which was financed by the large US food chain, Cumberland Farms. Newfoundland Energy took the necessary steps to re-activate the plant and through another Cumberland subsidiary, Newfoundland Processing, began shipping refined products to market in 1987.

In August 1994, Vitol B.V. acquired the refinery and renamed it North Atlantic Refining. The company undertook major renovations costing more than three hundred million dollars to improve efficiency and meet environmental standards. The refinery's capacity is 105,000 barrels of crude per day. The company's huge docking facility, the largest of its kind in North America – along with its more than seven million barrel storage capacity – gives it the capability of handling three-hundred-thousand-ton tankers. North Atlantic Refining employs approximately six hundred full-time workers at Come by Chance, as well as one hundred others in its marketing arm, North Atlantic Petroleum. The company exports over one million dollars worth of product annually.

WEST COAST OIL

Newfoundland's west coast region is the part of Newfoundland and Labrador with a geology most conducive to the presence of oil or natural gas. Most of the province consists of igneous and metamorphic structures, but the west coast, especially near the Port au Port peninsula, contains carbonate rocks, shale and other sandstone structures. These are the formations which have a potential for oil and gas deposits.

In the mid 1800s, oil seepage was observed at several west coast locations including St. Pauls's Inlet, Parsons Pond and Shoal Point. As early as 1867, John Silver, a Nova Scotian sawmill operator, heard local residents report oil slicks on Parsons Pond and began drilling on the north side of that location. The Parsons Pond area was part of the "French Shore," and following objections by France, which controlled the coastal area, drilling was stopped. It was not until 1894 that the Newfoundland Oil Company obtained rights to the area, and found oil. Between 1916 and 1926, an English firm named the General Oilfield Company operated three wells and a small refinery producing oil for local consumption; however, there was insufficient oil to make the operation economically viable. The short-lived Parsons Pond refinery was Newfoundland's only refinery until the 1960s.

In the 1990s, engineers and geoscientists, using modern seismic techniques, renewed exploration in the Port au Port area. Hunt Oil and PanCanadian spudded a well on the Port au Port Peninsula in 1994 which encountered hydrocarbon-bearing reservoirs. Hunt Oil and several partners also drilled in St. George's Bay in 1996, but with disappointing results. Several other companies have done limited drilling. As of 2002, Canadian Imperial Venture Corporation's Garden Hill well on the Port au Port Peninsula has produced some encouraging results and testing is ongoing.

GRAND BANKS EXPLORATION AND DEVELOPMENT

As previously mentioned, Newfoundland's geology, with the exception of pockets on the island's west coast, does not lend itself to the presence of recoverable petroleum. However, east of the island of Newfoundland, out towards the limit of Canada's economic offshore zone on the Grand Banks, the picture suddenly changes. This area, along with other segments off the east coast of Newfoundland and Labrador, possesses a number of pockets of sedimentary formation which contain oil and natural gas. This was realized by the major oil companies – particularly Mobil Oil, Imperial Oil Limited and Pan American Petroleum Corporation – who in the 1960s and 1970s began large scale seismic surveys of the area. The first survey was conducted in 1965, and by the following year, the first exploratory well was drilled. More wells were subsequently drilled at various sites off the east coast of Newfoundland and Labrador, followed by others near the edge of the Grand Banks.

After a period of disagreement between the provincial and federal governments over jurisdiction of offshore resources, an agreement was finally reached, and by the end of the 1970s drilling for oil began in earnest. Although oil was found in several exploratory wells in the 1970s, their flow rates did not warrant further development.

HIBERNIA

In September 1979 Chevron announced the first results from its test wells at the Jeanne d'Arc basin, in an area 315 kilometres east southeast of St. John's, designated as "Hibernia." The initial tests revealed a flow of eight hundred barrels a day, higher than that of any other well in the area. Further tests indicated a flow of more than 2600 barrels a day, which increased later to more than 12,500 barrels a day. Subsequent tests in the 1980s showed that the Hibernia area contained more than 615 million barrels of recoverable oil and 3.5 trillion cubic feet of natural gas.

Offshore drilling on Newfoundland's Grand Banks was not entirely without risks. One of the early semi- submersible rigs drilling off the east coast was the *Ocean Ranger*, the largest in the world at the time. This rig capsized in a fierce storm on February 15, 1982, with the loss of all eighty-four men on board.

It was not until 1990 that an agreement was reached between the oil companies and federal and provincial governments for development of the site. One of the major engineering issues was the type of production facility – a floating platform or a gravity-based structure. After considerable engineering analysis it was decided that the production platform at Hibernia would be gravity-based, and would sit on the ocean floor in about eighty metres of water. In September 1990, the first Hibernia platform design contract was awarded to Newfoundland Offshore Development Construction (NODECO), and the detailed design contract was subcontracted to Doris Development Canada (DDC). The steel-reinforced concrete Gravity Base Structure (GBS) was to be built at Bull Arm in Trinity Bay. Coincidentally, Bull Arm was also the site of another notable engineering happening almost 140 years earlier – the landing of the first transatlantic telegraph cable in 1858.

One of the biggest concerns of employing a GBS structure was its vulnerability to a collision with ice. To provide for this contingency, engineers designed the structure to withstand a hit by an iceberg weighing six million tons. To provide this level of protection, a fifteen metre

Challenge and Change — 207 — Oil and Gas

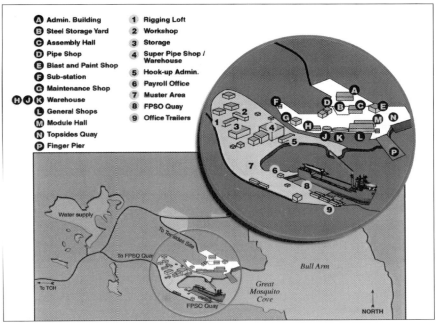

Bull Arm site map (Courtesy of HMDC)

Bull Arm fabrication site (Courtesy of HMDC)

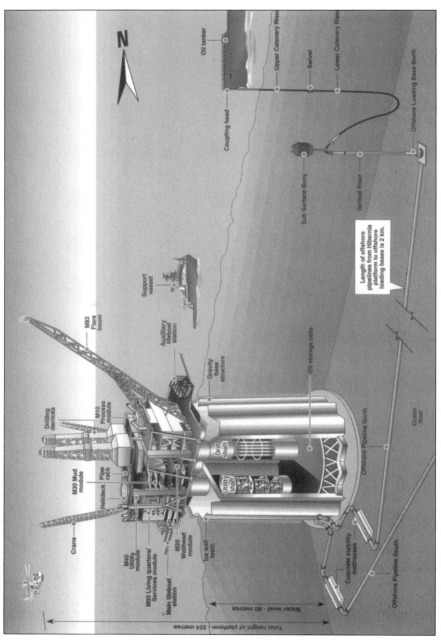

View of the Hibernia platform with its underwater components (Courtesy of HMDC)

thick concrete ice-wall was incorporated into the design. Inside the concrete outer shell of the GBS was an oil storage system capable of storing 1.3 million barrels of crude.

The topside of the drilling platform was made up of five "super modules," consisting of the process, wellhead, mud, utilities and accommodation components. The wellhead module, and several other topsides structures (including the flareboom, helideck, main and auxiliary lifeboat stations) were fabricated at Bull Arm. Two of the other super modules were fabricated in Korea and the remaining two in Italy.

The Bull Arm location was ideally located for oil platform construction as it was surrounded by steep hills to protect it from the wind, with a deep water construction site nearby. The facility was equipped with mechanical and electrical shops, wharfage, office, and accommodation space to tackle large scale projects. The Hibernia project was Bull Arm's first, followed shortly afterwards by the Terra Nova project.

In August 1994, pouring of concrete began for the GBS. The engineers decided on a slipforming process, whereby the concrete was continuously poured while the forms were slipped up the steel reinforcement. By October of that year, the concrete work was completed and the

Construction of the Hibernia GBS at Bull Arm (Courtesy of HMDC)

dry dock was flooded, successfully floating the structure. The 111 metre high structure used 450,000 tonnes of concrete along with 100,000 tonnes of reinforcing iron rebar.

The following month the structure was towed to the deepwater part of the harbour at Bull Arm where its tanks were flooded until the top of the GBS was just above the water surface. Meanwhile the

> **Bull Arm**
> The construction of the GBS at Bull Arm required many support services. Feeding the staff typically required on a monthly basis:
> 13,000 dozen eggs
> 16,000 pounds of potatoes
> 19,000 steaks
> 28,000 loaves of bread
> 30,000 litres of milk
> 2500 cases of fruit.

super modules were arriving from Korea and Italy and were mated with the wellhead unit on massive pontoon type barges. On February 27, 1997, tugs began towing the barges, with the 37,000 tonne topsides assembly onboard, out to the GBS, which was almost totally submerged. The barges were positioned over the top of the GBS, and once water was pumped out, the platform was allowed to rise to mate with the topsides unit, producing a structure 224 metres high.

Once testing was finished, on May 23, 1997, nine sea-going tugs began towing the completed platform to its final position, 315 kilometres offshore. This process took several days to complete. When it was in posi-

The **Ocean Ranger** sank in a storm February 15, 1982, with a loss of eighty-four men. (Courtesy of **The Express**)

> **Hibernia Quick Facts**
>
> *One of the Hibernia wells set a Canadian record when it tested a flow rate of 56,000 barrels of oil per day.*
>
> *The Hibernia project, at the time of construction, was the most expensive project undertaken in Canada.*
>
> *The completed weight of the Hibernia platform is approximately 1.2 million tonnes.*
>
> *The height of the Hibernia platform is more than half the height of New York's Empire State Building.*
>
> *Hibernia in 2002 was producing approximately five million barrels of oil per month.*

tion, the ballast tanks were flooded and on June 5, the platform settled to its final location. Tankers brought more than four hundred thousand tons of dry ballast to the platform to fill its interior tanks. By the end of the month, the operation was completed and the platform was firmly seated on the ocean floor. Its completed weight was almost 1.2 million tonnes. On July 28, 1997, the platform began drilling its first well, and obtained oil on November 17.

The Hibernia Gravity Base Platform is capable of drilling two wells at a time. The rig employs extended reach drilling technology which enables it to drill to about nine thousand metres vertically and seven thousand metres horizontally away from the platform. As of September 2002, seventeen oil producing wells had been drilled: fourteen in the Hibernia structure (thirty-seven hundred metres deep) and three in the Avalon/Ben Nevis structure (twenty-four hundred metres deep). It is expected that approximately eighty wells will be drilled over the twenty year life of the project. These wells produce 140,000 barrels of oil per day. As of August 2002, the estimate of recoverable reserve at the Hibernia field was 884 million barrels.

A major component of the Hibernia operation is the Offshore Loading System. This system employs sub-sea pipelines from the platform to the loading point, a sub-surface buoy, and loading hoses to the shuttle tankers. The loading point is located about two miles from the Hibernia GBS to minimize the risk of an accidental collision between the tankers and the rig. Oil is then offloaded to one of three 850,000 barrel shuttle vessels, the *Kometik*, *Mattea* or *Vinland*, which were specially constructed for transporting oil to the transshipment facility at Whiffen Head, Placentia Bay.

The owners of the Hibernia operation are Mobil Oil Canada Properties (33.125%), Chevron Canada Resources (26.875%), Petro-

*Towing the Hibernia platform to the Grand Banks (Courtesy of **The Telegram**)*

Canada Hibernia Partnership (25%), Murphy Atlantic Offshore Oil Company (6.5%), and Canada Hibernia Holding Corporation (8.5%).

TERRA NOVA

The Terra Nova Oilfield was discovered by Petro-Canada in 1984. It is also located in the Jeanne d'Arc basin, approximately 350 kilometres east-southeast of St. John's, and about thirty-five kilometres southeast of the Hibernia platform. The water depth is about ninety-five metres. The structure is made up of three fault blocks, the Graben, East Flank, and the Far East. The estimated recoverable reserves of these areas is approximately 406 million barrels.

The owners of the Terra Nova operation are Petro-Canada (33.99%), ExxonMobil Canada properties (22.00%), Norsk Hydro Canada Oil and Gas (15.00%), Husky Oil Operations Ltd. (12.51%), Murphy Oil Company Ltd. (12.00%), Mosbacher Operating Ltd. (3.5%), and Chevron Canada Resources (1.00%). The pre-production capital cost of the development was approximately $2.8 billion.

After extensive engineering study on different means of recovering the Terra Nova oil, the engineers decided to proceed with a Floating Production Storage and Offloading (FPSO) system, rather than a gravity based system such as that used at Hibernia. The FPSO vessel would use five thrusters and a dynamic positioning system to keep it stationary. The positioning system receives precise information from two global positioning satellites as well as from six transponders placed on the ocean floor.

Though Petro-Canada is the main owner and operator, many other companies shared responsibilities for the construction phase of the project. Petro-Canada was responsible for the shuttle tankers and drilling unit; CoFlexip Stena Offshore Newfoundland for the risers, flow line and glory holes; Agra Brown and Root for the process modules and FPSO vessel; FMC Offshore for the topsides turret and mooring system; Haliburton Energy and Services for the drilling unit; and Doris Con Pro for various aspects of the subsea development.

Development drilling was started at the Terra Nova oilfield by the semi-submersible *Glomar Grand Banks* in July 1999; this rig was replaced by the *Henry Goodrich* in February 2000. As of June 2002, ten wells had been drilled, six for the extraction of oil, and the others for gas and water injection. It is expected that twenty-four wells will be drilled over the life of the project.

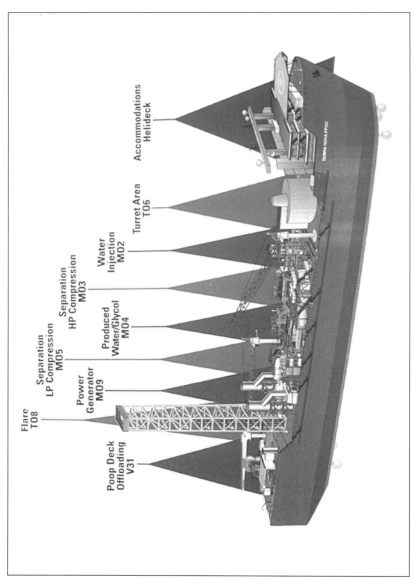

The Terra Nova FPSO, highlighting its various modules (Courtesy of Petro-Canada)

For the first six years of production, oil will be recovered at a rate of approximately 125,000 barrels per day from the Graben and East Flank areas.

The Terra Nova FPSO was fabricated in South Korea by Daewoo Heavy Industries, with work commencing in August 1998. The double hulled vessel, which can accommodate a crew of eighty, is 292.2 metres long and 45.5 metres wide, with an overall displacement of 196,000 tonnes. It has a crude oil storage capacity of 960,000 barrels and can handle 125,000 barrels a day. The vessel was officially named *Terra Nova FPSO* and left Korea on March 15, 2000, arriving at Bull Arm in June. Shortly afterwards, installation of the topsides components began. These included the Water Injection Module (1086 tonnes), the Power Generator module (1484 tonnes), the Flare Tower module (600 tonnes), and the Produced Water/Glycol module (1400 tonnes), all of which were fabricated at Bull Arm. The modules constructed outside the province – the Separation HP Compression module (2167 tonnes), and the Separation LP Compression module – were fabricated at Ardersier, Scotland.

One of the main sections of the FPSO is the seventy metre high and four thousand tonne turret, which is used for mooring as well as for connecting to the subsea systems. The FPSO swivels around this component, allowing the vessel to always face into the wind. The turret is dis-

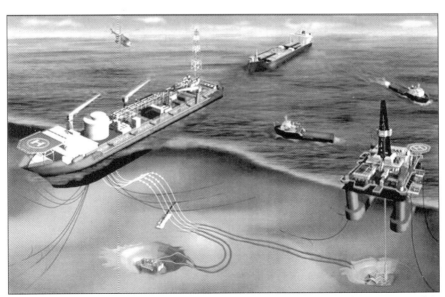

The various components of the Terra Nova Project (Courtesy of Petro-Canada)

> **Quick Terra Nova facts**
>
> Approximately $700 million was spent in Newfoundland on the design and construction phases of the Terra Nova project in Newfoundland.
>
> The Terra Nova FPSO is the first constructed to meet the severe weather conditions encountered off Newfoundland's coast.
>
> The FPSO's disconnectable turret, seventy metres high and weighing 4000 tonnes, is the world's largest.
>
> The Terra Nova field has estimated recoverable reserves of 370 million barrels of oil, and the year-end 2002 production target is 125,000 barrels a day.

connectable, which means the ship could disengage from the subsea components and head to a safe area if threatened by icebergs.

The crude passes through the turret and is sent to various processing modules where it is separated into oil, water and gas. The oil is then stored in the vessel's fourteen storage tanks, the largest having a capacity of approximately 78,000 barrels. Once the oil is ready to be unloaded, a shuttle tanker approaches the rig and a mooring line and twenty inch hose are put in place. Oil is pumped to the tanker at a rate of 50,000 barrels per hour. During transfer, water is pumped into ballast tanks to maintain the FPSO at its operating draught of 12.5 to 18.5 metres.

The recovered water is treated and dumped in the ocean. The gas is either re-injected into the well to help recovery or used to power equipment on the FPSO such as the gas turbine generators which produce the vessel's electricity. The FPSO's power generators can produce up to forty megawatts of electricity, enough to power a small city containing 50,000 homes. As a last resort, any excess gas is burned off through the flare tower. The vessel also has a system to convert sea water into drinking water,

> **Newfoundland and Labrador oil production**
>
> | Year | Barrels |
> |------|---------|
> | 1998 | 23799 |
> | 1999 | 36392 |
> | 2000 | 52798 |
> | 2001 | 54288 |
> | 2002 | 30950* |
>
> * Production to the end of April, 2002. Prior to 2002, only Hibernia oil production is reflected; in 2002, Terra Nova oil is also included.
>
> Source: Canada-Newfoundland Offshore Petroleum Board, May 2002

The Terra Nova FPSO at Bull Arm preparing for its voyage to the Grand Banks (Courtesy of Petro-Canada)

which can produce fifty cubic metres a day, more than twice *Terra Nova FPSO's* needs.

After testing at Bull Arm and sea trials in Trinity Bay, the FPSO left Bull Arm late in the night of August 2, 2001 and arrived at its position at the Terra Nova oil field on August 4. Oil production commenced in January 2002.

WHIFFEN HEAD TRANSSHIPMENT FACILITY

An essential requirement for a viable oil industry in the province is the ability to get the product to market. To achieve this, a transshipment facility was required to serve as a temporary storage area for oil delivered from the offshore rigs before it is loaded on ocean-going tankers. The ideal location was Whiffen Head, Placentia Bay, where Newfoundland Transshipment Limited (NTL) constructed a world-class facility. The site is equipped with large storage tanks and docking facilities capable of handling both shuttle as well as ocean-going tankers up to 159,000 dwt. In 2002, there was storage capacity for 2.5 billion barrels of crude on the site, which could be further expanded. The transshipment facility is served by the *Placentia Pride* and the *Placentia Hope*, two 5600-horse-

Transshipment facility at Whiffen Head, Placentia Bay
Some features of this facility are a 80,000 barrel per hour receiving capacity, 50,000 barrel per hour loading rate, 159,000 dwt tanker docking accommodation, and six storage tanks (expandable to ten), each with a working capacity of 500,000 barrels. (Courtesy of Newfoundland Transshipment Ltd.)

power, thirty-eight metre long escort and multi-purpose tugs which were constructed for NTL at the Friede Goldman shipyard at Marystown.

FUTURE OIL DEVELOPMENTS

As of 2002, only the Hibernia and Terra Nova oil fields were producing oil on the Grand Banks. The next field scheduled for development is White Rose, which is adjacent to the Hibernia and Terra Nova fields with a recoverable oil reserve of approximately 283 million barrels. This project will also use FPSO technology, but at 270 metres, the vessel will be slightly smaller than the Terra Nova unit. The Newfoundland government has given the go-ahead to Husky Oil to start development. It is expected that the White Rose project will create an annual employment of 350 to 375 over its four to five year construction phase until it goes on stream around 2005. The front end engineering design was awarded to Maersk/Sealand Engineering of St. John's. The topsides fabrication contract was awarded to Aker Maritime Kiewit Contractors, which will do much of the construction at the Marystown Shipyard, purchased earlier in 2002 by Peter Kiewit Sons. All of the engineering for the topsides will be done in Newfoundland and Labrador. When fully developed, the White Rose project is expected to produce between 75,000 to 110,000 barrels of oil per day.

Other oil fields in the area include Hebron (325 million barrels), Ben Nevis (55 million barrels), and West Ben Nevis (34 million barrels). These fields are within twenty kilometres of each other, and are currently being considered for development.

Memorial University of Newfoundland St. John's campus

Chapter Ten

Engineering and Geoscience Education and Research in the Province

The first school in Newfoundland was established in 1723 at Bonavista by the Society for the Propagation of the Gospel. Over the years, other schools were established throughout the colony, most of which were operated by various church denominations and religious orders with financial support from families. It was not until 1832 that the government started to play a role in delivering education services by providing financial assistance. In 1885, a public examination system was implemented. The largest schools were located in St. John's. They included Bishop Feild College, established in 1844, and re-located to its Bond Street location in 1928; St. Bonaventure's College, opened in the mid 1850s; and Prince of Wales College, established in 1919. These schools were among those providing the highest levels of education in Newfoundland until the establishment of Memorial College.

MEMORIAL UNIVERSITY OF NEWFOUNDLAND

Memorial University College (College) was founded in 1925 at the Parade Street campus in St. John's as a memorial to Newfoundlanders

The S. J. Carew Building
Opened in 1975, the S. J. Carew Building houses Memorial University's Faculty of Engineering and Applied Science.

who gave their lives in World War One. Its first president was John Lewis Paton, who servided in that capacity until 1933, when he was succeeded by Albert G. Hatcher. The College provided mainly two year courses (three years for pre-engineering) and served as a feeder school for mainland universities.

In 1930, the College began offering a structured three year engineering diploma program in association with the Nova Scotia Technical College (now part of Dalhousie University), which recognized the program as the first three years towards an engineering degree. In the program's inaugural year, Dr. Thomas H. Winter was the only engineering professor, who supervised thirteen students. Some of the students of this class were Dr. J. B. Angel, Ralph Higgins, Clarence Knight, R. V. Moores Dr. Clarence Powell, and Walter Woolfrey. C. A. D. MacIntosh replaced Dr. Winter the following year. In 1934, Dr. G. A. Frecker took over the engineering teaching duties and was followed by S. J. Hayes, who served from 1935 to 1941. In 1941, Stanley Carew replaced Professor Hayes as the sole engineering professor until John M. Facey joined the faculty in 1946.

The Alexander Murray Building
Built in 1990, the Murray Building houses Memorial University's Earth Sciences Department and the Centre for Earth Resources Research.

In 1949 the College became a full-fledged university and was renamed Memorial

University of Newfound-land (Memorial). The Engineering Department became part of the Faculty of Applied Science, with Professor Stanley J. Carew its first dean. Another staff member joined in 1955, and by 1966 there were six faculty members teaching slightly more than four hundred engineering students. The faculty has expanded considerably over the years: in 2002, the Faculty of Engineering and Applied Science contained forty-five instructors and almost ninety support staff.

Memorial University Engineering Faculty in the early 1960s
From the left, Howard Dyer, Anthony Nemec, Stanley Carew (Dean), and John M. Facey (Courtesy of A. C. Hunter Library, Newfoundland Collection)

For many years, after completing an engineering diploma at Memorial, candidates would qualify for a Bachelor's degree in electrical, mining, civil or mechanical engineering after two years of successful study at the Nova Scotia Technical College. This arrangement continued until September 1969, when Memorial began its own degree-granting engineering program. In the same year, a Masters of Engineering program was also implemented, followed two years later by a PhD. program in ocean engineering. In 1975, a new engineering building, named after S. J. Carew, was opened to house the engineering, forestry and earth sciences disciplines.

*MUN Deans of Engineering
From the top: Dr. S. J. Carew (1949-69), Dr. A. Bruneau (1969-74), and Dr. R. Dempster (1975-80)
(Courtesy of MUN and NP)*

The first Dean of Engineering and Applied Science after Memorial began conferring Engineering degrees was Dr. Angus Bruneau, formerly Director of General Engineering at the University of Waterloo. Dr. Bruneau served as dean between 1969 and 1974, and after a period in private practice was appointed President and Chief Executive Officer of Newfoundland Light and Power in 1986.

The curriculum at Memorial was designed around the concept of co-operative engineering: in addition to their regular academic regimen, students were also required to serve a significant amount of their training time working in an industrial environment. Memorial was one of the earliest universities to implement this approach, which has since been adopted in most other Canadian universities. Memorial's first electrical, mechanical and civil engineers graduated in 1974. There were seventy-five males in that group. The following year, Hillary Dawson became the first female to graduate from Memorial with a degree in engineering.

Over the years, Masters and PhD. graduate programs were also implemented in various engineering disciplines. Ocean and naval architecture engineering was introduced in 1982, and computer engineering in 1991; there are strong research programs in all these areas. In 2002, Memorial's Faculty of Applied Science and Engineering's enrollment included 1200 undergraduate, 118 Masters, and 38 PhD. students.

The Earth Sciences Department at Memorial is one of the largest in Canada, and offers one of country's most diverse earth sciences programs. Its main facility is the Alexander Murray building, which also houses the Centre for Earth Resources Research (CERR) and its extensive array of laboratory and testing equipment. Retired professor Dr. Harold "Hank" Williams became Memorial's first earth sciences graduate when he received a B.Sc. in Geology in May 1956. As of the fall of 2002, the department has a faculty of twenty-six, with fifteen support staff, and an enrollment of approximately eight hundred undergraduate and fifty graduate students. Dr. James Wright is currently Head of the department and Director of CERR.

Memorial has a number of engineering-related research centres, centres of excellence, and institutes. These include the Centre for Cold Ocean Resources Engineering (C-CORE), the Centre for Earth Resources, the Ocean Engineering Research Centre (OERC), and the Marine Institute. The Institute for Marine Dynamics of the National Research Council of Canada, while not part of Memorial, is adjacent to the campus and has a close working relationship with the university.

CENTRE FOR COLD OCEAN RESOURCES ENGINEERING (C-CORE)

The Centre for Cold Ocean Resources Engineering is a not-for-profit corporation of Memorial University which specializes in research relating to Canada's ocean resources, including the study of ice and icebergs, remote sensing, seabed physics, geotechnical modelling, space systems and applications, and intelligent systems. C-CORE was established in April 1975 with Harold Snyder as its first Director. Snyder had previously been a Vice-President with Brinco Ltd. in charge of construction at the Churchill Falls development. C-CORE depends on outside sources, including gov-

MUN Deans of Engineering
From the top: Dr. C. DiCenzo (1980-82), Dr. G. R. Peters (1982-92), and Dr. R. Seshradi (1993-present) (Courtesy of MUN Photographic Services)

ernment, industry and various funding agencies, for most of its funding for salaries and equipment. Memorial University remains responsible for general expenses such as accounting, personnel and utility services.

CENTRE FOR EARTH RESOURCES RESEARCH

The Centre for Earth Resources Research is part of the Department of Earth Sciences and is a research centre for earth sciences, including oil and gas. It was opened in 1990 in the Alexander Murray Building at Memorial University. CERR receives its research funding from sources outside the university, relying primarily on the oil and mining industries. Some of its major projects include seismic studies on the Port au Port peninsula and various site surveys for Newfoundland's offshore oil industry.

OCEAN ENGINEERING RESEARCH CENTRE (OERC)

The Ocean Engineering Research Centre is part of the Faculty of Engineering and Applied Science. Some of its areas of research expertise include ice-mechanics, marine structures and propulsion, underwater vehicles, vessel dynamics, motion simulation, sea ice mechanics, marine hydrodynamics and wave-structure interactions. OERC works closely with the National Research Council Institute for Marine Dynamics and has access to the Institute's facilities.

FISHERIES AND MARINE INSTITUTE

In 1964, the growth in educational facilities continued when the College of Fisheries, Navigation, Marine Engineering and Electronics (Fisheries College) was set up in St. John's at the old Memorial College campus on Parade Street. Its first president was Dr. William Hampton and in its first year of operation, 1464 students were enrolled. The Fisheries College was primarily established to offer technical and vocational training for the fishing industry, but graduates were not limited and could pursue other careers of their choosing. Vocational courses, which covered areas such as cooking, carpentry and rigging, were usually of a short duration, in the order of two weeks to several months. The technical courses were normally two or three years long, and students graduated with a Diploma of Technology. Nautical Science and Naval Architecture courses of various durations were also given to prepare students for

Fisheries and Marine Institute - Memorial University of Newfoundland

marine certificates for the Merchant marine, Coast Guard, fishing and shipbuilding industries.

Since 2001, the Institute has been known as the Fisheries and Marine Institute of Memorial University of Newfoundland. Courses are offered in support of the fishery, marine and petroleum industries, with diploma courses as well as degrees at the Baccalaureate and Masters level. In 2001, there were more than forty-five hundred students and one hundred thirty faculty.

INSTITUTE FOR MARINE DYNAMICS

Adjacent to Memorial University lies the National Research Council Institute for Marine Dynamics. This facility was established in 1985 as a national centre for research and development of ocean technology. The Institute collaborates with both Canadian and international universities, industries, and research agencies with the view of bringing new technology to Canada.

The Institute for Marine Dynamic's world class facilities include a seventy-five by thirty-two metre offshore engineering basin equipped

The two-hundred-metre-long towing tank at NRC's Institute for Marine Dynamics in St. John's (Courtesy of NRC-IMD)

with a current generation system and 168 wavemakers capable of producing a wide variety of wave types. There is also a two hundred metre long towing tank which likewise has wave-making capability as well as a ninety metre long ice tank, which is the longest in the world.

VOCATIONAL SCHOOLS

There was no organized vocational education system in Newfoundland prior to 1946. After World War Two, returning troops found that they had no or little or job preparation, having missed the opportunity to obtain a trade or skill. To help integrate these servicemen into the workforce, the government set up a Serviceman's Vocational school on the south side of St. John's, and offered courses in plumbing, sheet metal, electrical and several other trades. After Confederation with Canada in 1949, the school was also opened to civilians.

Realizing the importance of regulating trades in the province, the government implemented the Apprenticeship Act in 1953, which provided regulations for the major trades as well as for the mining and paper industries. In the early 1960s, the Canadian government announced a plan to supplement seventy-five per cent of the cost of construction of new trade and technical colleges. The Newfoundland government jumped at this opportunity and built a number of new facilities throughout the island. Prior to this announcement, the government had already decided to construct the College of Trades and Technology, which is located on Prince Philip Drive in St. John's. By the end of 1964, after an unprecedented expansion of vocational education, new colleges had been built at Bell Island, Burin, Carbonear, Clarenville, Conception Bay South, Corner Brook, Gander, Grand Falls, Lewisporte, Port aux Basques and Stephenville Crossing. In 1997, the vocational schools restructured and amalgamated to become the College of the North Atlantic (CONA). As of 2002, CONA operated seventeen campuses scattered across the island and Labrador, with the main campus at the Prince Philip Drive location, and offered seventy full time programs and three hundred part-time courses to approximately ten thousand students. Some of the programs offered include Applied Arts, Business, Health Sciences, Engineering Technology, Industrial Education/Trades, Information Technology, Natural Resources, Academic/Access programs and English as a Second Language.

Chapter Eleven

The History of APEGN

In 1948, the Engineering Institute of Canada (EIC) opened a branch in St. John's. The EIC was an association of Canadian engineers which shared professional technical information. Though Newfoundland was not yet part of Canada, the EIC invited its neighbouring Newfoundland engineers to join the association. E. Baillie became the branch's first president.

At that time, there was no regulatory body in Newfoundland to oversee the practice of professional engineering, and local EIC members became instrumental in setting up such an association. Consequently, the Association of Professional Engineers of Newfoundland (APEN) was established in 1952 with the purpose of regulating the practice of engineering in the province. The role of APEN was embodied by Newfoundland statute in the Newfoundland Engineering Profession Act, which was proclaimed in May, 1952.

The first general meeting of APEN was held on June 7, 1952 at the Memorial University campus on Parade Street in St. John's. Twenty-seven engineers were in attendance, and one of the first orders of business was the election of an executive.

Beverley Monkman, P. Eng. (1913-1999)
Monkman graduated with a degreee in Civil Engineering from the University of Alberta in 1940. In addition to being the president of APEN for its first two years, he also became president of APEG-GA in 1961. (Courtesy of APEGN)

Beverley A. Monkman was elected President and Ernest Dickinson was elected Vice-president. Monkman, who was born in Alberta, would later also become president of the Association of Professional Engineers, Geologists, and Geophysicists of Alberta. There were four councillors elected from the St. John's area, and four from outside the city. The former group consisted of Stanley J. Carew, M. A. Foley, Grant Jack, and Clarence Knight. The latter group contained Elmer L. Ball, H. Hughson, W. H. Maher, and William L. Stuewe.

The association's membership in its first year reached sixty-five, including four licensees and three engineers-in-training. The only surviving founding member in 2002 is Edward M. Martin, who prior to his retirement, worked with ASARCO in Buchans.

Notice
The Professional Engineers Association of Newfoundland

Notice is hereby given that from and after the expiration of one month from the date hereof, we, the undersigned, and others with us, all Members of the Engineering Institute of Canada, will make application to the Legislature of Newfoundland for the enactment of a Private Bill for the purpose of incorporating as a body corporate persons practising the profession of Engineers in the Province of Newfoundland with a view to the improvement of the Profession.

Comments and criticisms are invited from the public. Such comments and criticisms are to be addressed to the Clerk of the House of Assembly, St. John's, and are to be submitted to him not later than the 4th day of April, 1952.

Ernest Dickinson	**Wm. L. Stuewe**
Beverley Monkman	**Lex Myers**
Grant Jack	**Eric Hinton**
Brian Higgins	**Elmer L. Ball**
Stanley Carew	**Ed. Martin**

Newfoundland Gazetter - March 4, 1952

In 1953, APEN became a member of the Dominion Council of Professional Engineers, the forerunner of the Canadian Council of Professional Engineers (CCPE). In May, 1966, APEN began publishing the *Newfoundland and Labrador Engineer*, a regular newsletter, which highlighted engineering news and views of interest. This publication was succeeded by *Dialogue for Engineers* in 1978, and subsequently renamed *Dialogue for Engineers and Geoscientists* in 1990.

**Charter Members
of the
Association of Professional Engineers of Newfoundland
1952**

John B. Angel	H. Forbes Roberts
E. L. Baillie	Milton G. Green
J. Boyd Baird	H. J. Hermanson
Elmer L. Ball	Brian C. Higgins
W. L. Ball	Eric Hinton
Alex M. Butt	J. M. Hopkins
Stanley J. Carew	Grant R. Jack
Charles H. Conroy	Clarence A. Knight
D. L. Cooper	Ed Martin
George W. Cummings	Beverley A. Monkman
G. H. Desbarats	J. W. Morris
Ernest Dickinson	G. A. Myers
M. A. Foley	William L Stuewe

In 1968, the Association of Engineering Technicians and Technologists (AETTN) of Newfoundland was formed and became affiliated with APEN. One of this association's purposes was to improve the competence of technicians and technologists and to regulate standards for its members while protecting the public interest. For several years, APEN participated in a Board of Examiners to certify new AETTN members. In 1994, however, AETTN reorganized to become more autonomous and the two organizations dissolved their affiliation.

The Engineering Profession Act was modified in 1975 to provide for the regulation of engineering performed by partnerships, associations and limited companies. A further change was implemented in 1989, when the Act was replaced by the Engineers and Geoscientists Act, extending

Engineering Week at the Village Mall - One of the activities sponsored by APEGN (Courtesy of APEGN)

the association's authority to the regulation of geoscience. The association's name was thereby changed to the Association of Professional Engineers and Geoscientists of Newfoundland, and the first geoscientists were admitted in 1990.

There are currently four regional chapters of APEGN, which bring greater focus to engineering and geoscience in their respective areas. These sections are the Central (formed in 1970), Eastern (1972), Western (1971), and Labrador (1977).

Over the years APEGN has actively participated in community affairs by taking positions on engineering and geoscience matters, such as the Hibernia and the Voisey's Bay developments. It has also been active in education, and established an engineering scholarship at Memorial University in 1952. As of 2002, APEGN offers a number of scholarships and awards in engineering and geoscience to Memorial University students. APEGN also actively supports the employment of student engineers and geoscientists as part of their co-operative education program, and also participates in Engineering Week, an event encouraging young students to select engineering and geoscience as a career. In 1994, APEGN inaugurated Future SET (Future Scientists, Engineers, and

Technologists), a summer science camp at Memorial University for school children between grades three and eight.

As of 2002, the membership of APEGN consisted of 1682 Professional Engineers, 250 Engineers-in-Training, 176 Professional Geoscientists, 20 Geoscientists-in-Training and 163 Life Members. There were also 34 Licensees (non citizen or landed immigrant) and 9 Honourary members. There were also 442 Engineering and Geoscience firms registered with APEGN, 153 of which were based inside the province and 289 outside.

Over the past half century, female engineers have made great strides. In 1980, Ann Bridger became the first female to be registered as a Professional Engineer in Newfoundland. As of 2002, female engineers and geoscientists constituted a growing proportion of APEGN's membership, including seventy-eight Professional Engineers, forty-three Engineers-in-Training, twelve Professional Geoscientists and four Geologists-in-Training.

APEGN registrations (1997 - 2001)

Membership Category	1997	1998	1999	2000	2001
Honorary/Life Membership	143	149	147	156	165
Professional Engineers	1548	1570	1557	1591	1665
Professional Geoscientists	185	180	180	166	168
Dual P. Eng./P. Geo.	9	13	12	12	13
Engineers-in-Training	207	193	238	243	258
Geoscientists-in-Training	6	4	7	13	20
Sub-Total	2098	2109	2141	2181	2289
Restricted Licensees	15	3	1	1	1
Engineering Licensees	39	28	33	26	26
Geoscience Licensees	12	14	12	11	9
TOTAL	2164	2154	2187	2219	2325

Presidents of APEN and APEGN
1952 to 2002

B. A. Monkman P. Eng
1952-53

E. P. Dickinson P. Eng
1953-54

E. L. Ball P. Eng
1954-55

Carl A. Knight P. Eng
1955-56

H. B. Carter P. Eng.
1956-57

Stanley J. Carew P. Eng.
1957-58

M. G. Green P. Eng.
1958-59

F. Gover P. Eng.
1959-60

R. Wiseman P. Eng.
1960-61

C. H. Conroy P. Eng.
1961-62

F. D. Grant P. Eng.
1962-63

C. R. Vivian P. Eng.
1963-64

J. B. Angel P. Eng.
1964-65

G. N. Cater P. Eng.
1965-66

R. W. Myers P. Eng.
1966-67

P. V. Young P. Eng.
1967-68

K. F. Duggan P. Eng.
1968-69

W. W. Cossitt P. Eng.
1969-70

R. A. Kieley P. Eng.
1970-71

B. C. McGrath P. Eng.
1971-72

R. T. Parsons P. Eng.
1972-73

T. S. Goodyear P. Eng.
1973-74

M. T. O'Brien P. Eng.
1974-75

K. F. St. George P. Eng.
1975-76

H. J. Dyer P. Eng.
1976-77

A. R. Stanford P. Eng.
1977-78

J. G. Evans P. Eng.
1978-79

J. F. Green P. Eng.
1979-80

J. P. Henderson P. Eng.
1980-81

J. G. Kennedy P. Eng.
1981-82

G. R. Peters P. Eng.
1982-83

A. C. Green P. Eng.
1983-84

W. R. Haynes P. Eng.
1984-85

R. G. Scott P. Eng.
1985-86

E. K. Jerrett P. Eng.
1986-87

D. A. Dalley P. Eng.
1987-88

R. W. Scammell P. Eng.
1988-89

A. Edmunds P. Eng.
1989-90

S. M. McLean P. Eng.
1990-91

G. J. Follett P. Eng.
1991-92

R. A. Squires P. Eng.
1992-93

D. R. Finch P. Eng.
1993-94

W. J. Newton P. Eng.
1994-95

S. D. Banfield P. Eng.
1995-96

G. C. Emberley P. Eng.
1996-97

D. Whalen P. Eng.
1997-98

G. A. Suek P. Eng.
1998-99

K. R. Gosse P. Eng.
1999-00

A. A. Dawe P. Eng.
2000-01

C. Sheppard P. Eng.
2001-02

Chapter Twelve

Epilogue

The history of engineering in Newfoundland has indeed been an exciting one. The province has had a number of significant firsts, and while many of the engineering projects carried out within Newfoundland and Labrador may have been routine, many have been of a world class stature. From the engineering feats of the early transatlantic cables in the 1800s, to the early days of Newfoundland's first railway, its mines at Bell Island and Buchans, its paper mills at Grand Falls and Corner Brook, its highway construction, and its oil development, engineering has been at the forefront.

The engineering achievements of the past, as would be expected, pale in comparison to the engineering and geoscience projects of recent years. The Churchill Falls development along with the Hibernia and Terra Nova projects are engineering and geoscience landmarks, not only from a provincial perspective, but also at the national and international levels.

As to the future of engineering and geoscience in the province, in early 2002 several large engineering projects were being contemplated. Firstly, there is the potential diversion of rivers in Quebec to allow an additional one thousand megawatts of power to be generated at Churchill Falls. This prospect has been quashed at the political level, but nevertheless may rise again. There is also the development of the Lower Churchill River, which still has an unutilized potential of almost three thousand megawatts of hydro-electric power. Another concept of interest is a tunnel beneath the Strait of Belle Isle, which would provide not only vehicular access between Labrador and Newfoundland, but also a corridor for the transmission of electrical power to the island part of the province. This project would cost approximately ten to twelve billion dollars. Financing such megaprojects has always been a challenge for Newfoundland and Labrador. However, when the right economic and business climate is reached for industry and government, the development of the unutilized power on the Churchill River will almost certainly go ahead.

In early 2002, the White Rose and Voisey's Bay developments were announced, and engineering and geoscience work is under way. White Rose, which is based on a FPSO facility, was briefly discussed in Chapter Nine. For this project, almost all the engineering and geoscience work for the topsides will, for the first time, be done in Newfoundland and Labrador. The White Rose project will cost $2.3 billion and oil production is slated to begin in 2005.

The Voisey's Bay project will be undertaken by the International Nickel Company of Canada (INCO), which owns the rights to a huge nickel deposit at Voisey's Bay, thirty-five kilometres south of Nain, Labrador. INCO and the Government of Newfoundland and Labrador have settled on a framework for an agreement which they expect to sign in September. At the time of this publication, INCO was planning to develop an open pit mine at Voisey's Bay as well as a hydro-metallurgical processing facility at Argentia, Placentia Bay. Over its estimated life of thirty years the project will cost 2.9 billion dollars.

This volume has reviewed most of the province's major engineering and geoscience projects; however, there are scores of others which could have been covered had space permitted. Engineers and geoscientists are rarely in the limelight at official openings of new highways, mills or oil rigs. Most people accept development as progress, and few recognize or understand the engineering and geoscience skills required to bring these projects to completion. The engineers and geoscientists behind major projects are seldom publicly recognized; this holds true tenfold for the engineers and geoscientists who mastermind the more routine projects, and who have also made significant contributions. It is hoped that by examining the major projects, this book has brought to light the contributions made by APEGN professionals to the development of the province.

Appendix

Award of Merit

The Award of Merit was established to recognize members of APEGN who have made outstanding contributions to the profession and/or to the community. This recognition takes the form of an "Award of Merit" scroll which is normally presented to the recipient at the Awards Dinner during the APEGN Annual Conference.

Recipients of The Award of Merit

1977 - Ernest Dickinson, P. Eng.
        Stanley Carew, P. Eng.
        John B. Angel, P. Eng.
1978 - C. Harry Conroy, P. Eng.
1979 - Clarence W. Powell, P. Eng.
        George P. Hobbs, P. Eng.
1980 - Angus A. Bruneau, P. Eng.
1981 -
1982 -
1983 - Howard Dyer, P. Eng.
        Harold Lundrigan, P. Eng.
1984 - Robert A. Robertson, P. Eng.
        Kevin St. George, P. Eng.
1985 - John G. Evans, P. Eng.
        Wallace Read, P. Eng.
1986 - G. Ross Peters, P. Eng.
1987 - Anthony Brait, P. Eng.
1988 - Roy Murley, P. Eng.
1989 - Roy Myers, P. Eng.
1990 - George N. Cater, P. Eng.
1991 - Eric Jerrett, P. Eng.

Dr. Rex Gibbons, P. Geo.
1992 - John M. C. Facey, P. Eng.
1993 - Don Wilson, P. Eng.
1994 - Dr. Azizur Rahman, P. Eng.
Ernest A. Langins, P. Eng.
1995 - Gordon W. Butler, P. Eng.
Eric Swanson, P. Geo.
Peter Young, P. Eng.
1996 - David Collett, P. Eng.
Dr. James Sharp, P. Eng.
Frank Grant, P. Eng.
1997 - Dr. Mahmoud Haddara, P. Eng.
1998 - George Desbarats, P. Eng.
John Fleming, P. Geo.
Frederick Vivian, P. Eng.
Rex Parsons, P. Eng.
1999 - Dr. Hugh Miller, P. Geo.
R. Frank Davis, P. Eng.
2000 - Thomas Kierans, P. Eng.

Honourary Membership

Non APEGN members are eligible to receive Honourary Membership in the Association as provided for in the by-laws. Honourary Membership in the Association may be granted by Council to a person either eminent in the profession of Engineering or Geoscience or outside of these professions who, in the opinion of Council has uniquely distinguished himself/herself, and who is not otherwise eligible for Life Membership.

Recipients of Honourary Membership:
1993 - Clarence Randell
Vince Withers
John Power

1995 - William Legge
1996 - Jim Cooper
1997 - John McNeill
1998 - Harvey Weir
1999 - Carolyn Emerson
       Judith Whittick

Award for Service

The Award for Service is awarded to members of APEGN who have served their profession diligently for many years and who have also made substantial contributions to the Association and to the advancement of the profession.

Recipients of the Award for Service:

1991 - Doug Squires, P. Eng.
1992 -
1993 - Howard Dyer, P. Eng.
1994 - Heber Bowering, P. Eng.
1995 - Rick Gosse, P. Eng.
       Doug Goodridge, P. Eng.
       Gerry Suek, P. Eng.
1996 - John Fahey, P. Eng.
       Don Finch, P. Eng.
       Gary Follett, P. Eng.
1997 - Dr. Hugh Miller, P. Geo.
1998 - Peter Nell, P. Eng.
       Roger Angel, P. Eng.
       Dr. Peter Smith, P. Eng.
1999 -
2000 - Darlene Whalen, P. Eng.
2001 - Samuel D. Banfield, P. Eng.

Environmental Award

The Environmental Award recognizes the application of science, technology and engineering to human and resource environmental management in Newfoundland and Labrador. Its purpose is to emphasize the use of science, technology, and modern organization techniques by the engineering profession and others to protect and enhance the environment.

Recipients of the Environmental Award:
1994 - Royal Oak Mines Ltd.
1995 - Hibernia Management & Development Company Ltd.
1996 - NewTel Communications Ltd.
1997 - Public Works & Government Services Canada
1998 - Corner Brook Pulp & Paper Ltd.
1998 - Iron Ore Company of Canada
1999 - North Atlantic Refinery Limited
2000 - Newfoundland Power
2001 - Metal World, Department of National Defense and Servco Environmental Services

Early Accomplishment Award

The APEGN Early Accomplishment Award is given to Association members in recognition of exceptional achievement in the early years of an engineer's or geoscientist's professional career. The member will have shown outstanding work-related achievement and notable leadership in the profession and /or community within ten years of eligibility for professional registration in Canada.

Recipients of the Early Accomplishment Award
1996 - David Newbury, P. Eng.
       Emad Rizalla, P. Eng.
       Anthony Whalen, P. Eng.
1997 - Dr. Daniel Walker, P. Eng.

1998 - Dr. Raymond Gosine, P. Eng.
1999 - Miguel Pazos, P. Eng.
2000 - Peter Reid, P. Eng.

Community Service Award

The APEGN Community Service Award is given to Association members in recognition of outstanding service and dedication to society. The award recognizes a substantial contribution to the well-being of society through community organizations, government sponsored activities or humanitarian work. The contribution may be made in one or more of the fields of social service, religion, politics, education, sports, recreation or the arts and may include either voluntary or paid service, elected or appointed. Sustained activity over an extended period of time is an asset.

Recipients of The Community Service Award
1996 - Neil Windsor, P. Eng.
       Terence Goodyear, P. Eng.
1997 - Kevin St. George, P. Eng.
       Roger Flood, P. Eng.
1998 - John Neville, P. Eng.
1999 - William Newton, P. Eng.
2000 -
2001 - Robert G. Scott, P. Eng.

Teaching Award

This award recognizes an exemplary contribution by an individual in the areas of engineering and/or geoscience education.

Recipients of the Teaching Award
1999 - Dr. Johannes Molgaard, P. Eng.
1999 - Professor Michael Bruce-Lockhart, P. Eng.
2000 - Dr. Neil Bose, P. Eng.
2001 - Dr. Leonard Lye, P. Eng.
2001 - Dr. Hesham Marzouk, P. Eng.

Registered Members of the Association of Professional Engineers and Geoscientists of Newfoundland as of September 2002

Abbiss, Harry J., P. Eng.
Abbott, Dean A.T., P. Eng.
Abbott, Stephen H., P. Eng.
Abbott, Nelson R., P. Eng.
Abdel-Razek, Dr. A. Kamal, P. Eng.
Abdulezer, Alfred, P. Eng.
Aboulazm, Azmy F., P. Eng.
Abrahamson, Carman Wayne, P. Eng.
Acharya, Vipinchandra, P. Eng.
Achtarides, Dr. T. A., P. Eng.
Acton, David, P. Eng.
Adam, Darcee, P. Eng.
Adams, Andrew John, P. Eng.
Adams, Christopher, P. Eng.
Adams, John G., P. Eng.
Adams, Calvin M., P. Eng.
Adams, Robert, P. Eng.
Alacoque, Hubert, P. Eng.
Alaverdy, Peter V.E., P. Eng.
Albrechtsons, Eric A., P. Geo.
Alcock, Paul H., P. Eng.
Alexander, Peter, P. Eng.
Alexander, Benedict F., P. Eng.
Allanson, Geoffrey W., P. Eng.
Allen, C. Philip, P. Eng.
Anderson, Kerrine Lee, P. Eng.
Anderson, Gregory H., P. Eng.
Anderson, William J., P. Geo.
Anderson, Paul, P. Eng.
Anderson, Walter J., P. Eng.
Andrews, Peter W., P. Eng./P. Geo.
Andrews, Kenneth C., P. Geo.
Andrews, Mervin G., P. Eng.
Andrews, R. John, P. Eng.
Andrews, Alan A., P. Eng.
Andrews, Newman L., P. Eng.
Angel, Roger F., P. Eng.
Antle, Gerard B., P. Eng.
Appleby, Richard C., P. Eng.
Arambarri, Eduardo, P. Eng.
Arbuckle, Trevor, P. Eng.
Argirov, Nikolay, P. Eng.
Argue, Larry W., P. Eng.
Argue, David Glen, P. Eng.
Arklie, Albert E., P. Eng.
Arsenault, Joseph A., P. Eng.
Arthur, Douglas J., P. Eng.
Ash, H. Wayne, P. Eng.
Ashton, Elham, P. Eng.
Atkinson, Ian Maxwell, P. Geo.
Attwood, Paul G., P. Eng.
Au, Garfield, P. Eng.
Augustyniak, Mieczyslaw, P. Eng.
Avery, Peter, P. Eng.
Avery, Roland I., P. Eng.
Axnick, Fred F., P. Eng.
Aylward, James F., P. Eng.
Aziz, Mohammed, P. Eng.

Babstock, Reginald A., P. Eng.
Bailey, Christopher, P. Eng.
Bailey, David K., P. Eng.
Bailey, Raymond T., P. Eng.
Baird, William P., P. Eng.
Bajzak, Denes, P. Eng.
Bak, Dennis J., P. Eng.
Baki, Emad Magdi, P. Eng.
Balbaa, Raouf H. M., P. Eng.
Balcombe, Gordon, P. Eng.
Ball, Shawn L.W., P. Eng.
Balodis, Martin Paul, P. Eng.
Banfield, Samuel D., P. Eng.
Barbour, Stephen A., P. Eng.
Barbour, David M., P. Geo.
Barbour, Bruce, P. Eng.
Barbour, Les, P. Eng.
Barclay, Anthony, P. Eng.
Barnard, Joanna Mary, P. Eng.
Barnes, Marvin W., P. Eng.
Barnes, Robert D., P. Eng.
Barnes, Brian E., P. Eng.
Barnes, K. Jane, P. Eng.
Barrett, Perry R., P. Eng.
Barrett, Kelvin P., P. Eng.
Barrington, Michael A., P. Eng.
Barron, J. Leonard, P. Eng.
Barry, Vanessa L., P. Eng.
Barry, John P., P. Eng.
Barthe, Pierre, P. Eng.
Bartlett, Kenneth D., P. Eng.
Bartlett, Ralph, P. Eng.
Basha, Michael G., P. Geo./P. Eng.
Baskanderi, Habib, P. Eng.
Bastow, Philip, P. Eng.
Batstone, Paul S., P. Eng.
Batterson, Martin, P. Geo.
Bauld, Bruce R., P. Eng.
Bavis, Kirk A., P. Eng.
Baxter, Dennis M., P. Eng.
Bazeley, David, P. Eng.
Beaton, Jefferey B., P. Eng.
Beattie, Borden W., P. Eng.
Beaulieu, Gerard R., P. Eng.
Beaulieu, Christine, P. Eng.
Becker, Norbert K., P. Eng.
Bedford Jr., Clay P., P. Eng.
Beersing, Anil, P. Eng.
Behie, Stewart W., P. Eng.
Bell, William H., P. Eng.
Bell, Robert C., P. Geo.
Bell, Roderick A., P. Eng.
Bennett, E.w. Scott, P. Eng.
Bennett, Gilbert J., P. Eng.
Bennett, Maxwell D., P. Eng.
Bennett, I. David, P. Eng.
Benson, Raymond P., P. Eng.
Benson, Tolson, P. Eng.

Benson, F. Darryl, P. Eng.
Bentley, Philip John, P. Eng.
Beresford, Edward J., P. Eng.
Beresford, Thomas P., P. Eng.
Bergeron, Luc, P. Eng.
Berthelot, Eric Michael, P. Eng.
Berthiaume, Guy, P. Eng.
Bertolo, Daniel John, P. Eng.
Besaw, Robert M., P. Eng.
Beshay, Mamdouh Z.H., P. Eng.
Best, William W., P. Eng.
Bettney, Mark, P. Eng.
Beverley, Curtis G., P. Eng.
Bezanson, Kevin K., P. Eng.
Bharj, Tarlochan S., P. Eng.
Billard, David G.F., P. Eng.
Billing, Darryl C., P. Eng.
Bishop, Randolph G., P. Eng.
Bishop, Gary Alston, P. Eng.
Bishop, Gerald Owen, P. Eng.
Bisson, Martin, P. Eng.
Blackmer, Andrew, P. Geo.
Blackmore, Eliol Jr., P. Eng.
Blackmore, Michael F., P. Eng.
Blackwood, R. Frank, P. Geo.
Blake, Mark, P. Geo.
Blondin, Michel, P. Eng.
Blundon, Daniel M., P. Eng.
Bobby, Walter, P. Eng.
Bohaker, Brian R., P. Eng.
Boivin, Richard D., P. Eng.
Boland, Barry Keith, P. Eng.
Boland, Jeffrey Leo, P. Eng.
Bonner, G. Douglas, P. Eng.
Boone, Lorne C., P. Eng./P. Geo.
Boone, Aldwin H., P. Eng.
Boone, E. Keith, P. Eng.
Booton, Michael, P. Eng.
Borek, Aifons Stefa, P. Eng.
Bose, Neil, P. Eng.
Bosnjak, Tomislav, P. Eng.
Boudreau, Eric, P. Eng.
Bouhamdani, Mustapha, P. Eng.
Bourden, David, P. Eng.
Boutet, Robert, P. Eng.
Bouzane, Albert G., P. Eng.
Bowden, David Keith, P. Eng.
Bowering, Heber, P. Eng.
Bowering, E. David, P. Eng.
Bown, Harry Gordon, P. Eng.
Boyce, W. Douglas, P. Geo.
Boyle, Albert G., P. Eng.
Brace, Terry, P. Geo.
Bradbury, William John, P. Eng.
Bradley, Pearce W., P. Geo.
Bradley, Robert, P. Eng.
Bradshaw, Gary G., P. Eng.
Bradshaw, Janet L., P. Eng.

Bragg, Donald Mark, P. Eng.
Bragg, Daniel J., P. Geo.
Brake, John H., P. Eng.
Brake, Randy, P. Eng.
Branton, Robert, P. Eng.
Breen, Gordon, P. Eng.
Brennan, Gerald J., P. Eng.
Brennan, Paula, P. Eng.
Brennan, Donald P., P. Eng.
Brennan, Bernard E., P. Eng.
Breton, Dany Paul, P. Eng.
Breton, Rejean M., P. Eng.
Brewer, Kevin J., P. Geo.
Bridger, Barry, P. Eng.
Bridger, Ann E., P. Eng.
Briffett, Michael, P. Eng.
Brinston, Tony, P. Eng.
Brisson, Maurice, P. Eng.
Brockbank, William J., P. Eng.
Brockerville, Blake G., P. Eng.
Broders, Paul P., P. Eng.
Brodie, Frederick N., P. Eng.
Brooke, Ronald E., P. Eng.
Brooker, Terry D., P. Eng.
Brooking, Donald C., P. Eng.
Brooks, Earnest A., P. Geo.
Brophy, Donald M., P. Eng.
Brown, Harold J., P. Eng.
Brown, Kevin S., P. Eng.
Brown, Murray W., P. Eng.
Brown, Derek L., P. Eng.
Brown, David H., P. Eng.
Brown, Henry, P. Eng.
Brown, Jack H., P. Eng.
Browne, Raymond J., P. Eng.
Browning, David A., P. Eng.
Bruce, Colin Scott, P. Geo.
Bruce, Kenneth, P. Eng.
Bruce, Gary, P. Eng.
Bruce, James R., P. Eng.
Bruce-Lockhart, Michael, P. Eng.
Bruckner, Michael G., P. Eng.
Bruneau, Dr. Stephen, P. Eng.
Bruneau, Angus A., P. Eng.
Buckley, Terry J., P. Eng.
Budgell, Hubert G., P. Eng.
Budgell, Wilbert S., P. Eng.
Budwill, Sven, P. Eng.
Bugden, F. Craig, P. Eng.
Bugden, Jeffrey M., P. Eng.
Buggie, William J., P. Eng.
Buglar, Lloyd P., Eng.
Burden, Arthur D., P. Eng.
Burden, Elliott, P. Geo.
Burden, Steven Lloyd, P. Eng.
Burden, R. Jack, P. Eng.
Burden, D. Joseph, P. Eng.
Burke, Stephen G., P. Eng.
Burnett, James, P. Eng.
Burry, Oral, P. Eng.
Burry, Anthony, P. Eng.
Bursey, Rodney G., P. Eng.
Bursey, Arthur G., P. Eng.
Bursey-Snow, Lisa, P. Eng.
Bursey, G. Glenn, P. Geo.
Bursey, Thomas E., P. Eng.

Burt, David G., P. Eng.
Burton, James C., P. Eng.
Burtt, David M., P. Eng.
Bussard, Robert G., P. Eng.
Butler, Lorne J., P. Eng.
Butler, Robert, P. Eng.
Butler, John J., P. Eng.
Butler, Myles Scott, P. Eng.
Butler, Daniel H., P. Eng.
Butler, David J., P. Geo.
Butler, Frederick D., P. Eng.
Butler, Gordon W., P. Eng.
Butt, Roger, P. Eng.
Butt, Edward D., P. Eng.
Butt, Kenneth A., P. Eng.
Butts, Floyd E., P. Eng.
Byrne, Jerome, P. Eng.
Byrne, Patrick J., P. Eng.
Byrne, Kevin, P. Eng.
Cadigan, John R., P. Eng.
Cahill, Moya N., P. Eng.
Cahill, Frederick J., P. Eng.
Cahill, John F., P. Eng.
Caines, Jack, P. Eng.
Calderon, Antonino, P. Eng.
Caldwell, William C., P. Eng.
Callahan, Timothy J., P. Eng.
Cameron, Donald J., P. Eng.
Cameron, Robert M., P. Eng.
Cameron, A.E. Marston, P. Eng.
Campbell, Steven, P.Eng.
Campbell, Gordon N., P. Eng.
Campbell, D. Scott, P. Eng.
Campbell, James E., P. Eng.
Carew, Paul S., P. Eng.
Carey, Edward I., P. Eng.
Carlson, John E.C., P. Eng.
Carnell, Geoffrey C., P. Eng.
Carnell, John J.M., P. Eng.
Carpenter, Andrew R., P. Eng.
Carrick, Gordon J., P. Eng.
Carrigan, Sean, P. Eng.
Carroll, William J., P. Eng.
Carson, Gerald C., P. Eng.
Carter, Terence, P. Eng.
Carter, Gregory R., P. Eng.
Carter, Reginald G., P. Eng.
Casey, Jack William, P. Eng.
Cater, Robert, P. Eng.
Cater, Neil E., P. Eng.
Cater, George N., P. Eng.
Chabot, Daniel, P. Eng.
Chafe, Allan, P. Eng.
Chamois, Paul, P. Geo.
Champion, Rodney D., P. Eng.
Charamis, Panayotis, P. Eng.
Chari, Tuppal R., P. Eng.
Charron, Langis, P. Eng.
Chartrand, Claude, P. Eng.
Chaulk, Derrick Ward, P. Geo.
Chaulk, Neil, P. Eng.
Chaulk, B. Bradford, P. Eng.
Chaytor, Gregory T., P. Eng.
Chaytor, Stephen J.R., P. Eng.
Cheater, Brian James, P. Eng.
Cheeke, Deanna, P. Eng.

Cheema, Sukhminder, P. Eng.
Cheeseman, Arthur A., P. Eng.
Chen, Zhiwei Wayne, P. Eng.
Chen, Allan P., P. Eng.
Chevalier, Robert, P. Eng.
Chevarie, Camille J., P. Eng.
Chevarie, Jean-Maurice, P. Eng.
Ching, Chan Yu, P. Eng.
Chipman, Wayne I., P. Eng.
Chislett, Anthony, P. Eng.
Chung, George J., P. Eng.
Churchill, Rodney, P. Geo.
Churchill, Matthew M., P. Eng.
Churchill, Ivan R., P. Eng.
Churchill, Todd, P. Eng.
Churchill, N. Wayne, P. Eng.
Clark, William J., P. Eng.
Clark, John I., P. Eng.
Clark, Jody Patrick, P. Eng.
Clarke, B. Randolph, P. Eng.
Clarke, Clyde J., P. Eng.
Clarke, Bradley J., P. Eng.
Clarke, Johnny, P. Eng.
Clarke, Ingrid E., P. Eng.
Clarke, Peter M., P. Eng.
Clarke, Albert W., P. Eng.
Clarke, Wade L., P. Eng.
Clarke, Harold S., P. Eng.
Clarke, Geoffery R., P. Eng.
Cloutier, Maxime, P. Eng.
Clow, Trent David, P. Eng.
Coates, Howard James, P. Geo.
Coates, Gerald, P. Eng.
Coates, Terrance, P. Eng.
Cochrane, Augustine, P. Eng.
Cochrane, William C.J., P. Eng.
Coffin, Neal, P. Eng.
Colbourne, Overton A., P. Eng.
Cole, Gordon T., P. Eng.
Cole, Ronald A., P. Eng.
Cole, Brian W., P. Eng.
Cole, Rosanne E., P. Eng.
Cole, C. Brad, P. Eng.
Cole, Jason C., P. Eng.
Cole, Keith T., P. Eng.
Cole, Leo J., P. Eng.
Collett, David T., P. Eng.
Collins, Timothy S., P. Eng.
Collins, Celestine, P. Geo.
Collins, George, P. Eng.
Collins, R. Wayne, P. Eng.
Collins, Stephen F., P. Eng.
Collins, Michael, P. Eng.
Collins, Spencer H., P. Eng.
Collins, William T., P. Eng.
Collins, Christopher, P. Geo.
Collins, Wilson S., P. Eng.
Colman-Sadd, Stephen P., P. Geo.
Combden, J. Glenn, P. Eng.
Connolly, Daniel, P. Eng.
Connors, Robert C., P. Eng.
Connors, Tom, P. Eng.
Conroy, James F., P. Eng.
Cook, Norman B., P. Eng.
Cooke, Lloyd S., P. Eng.
Coombs, Bruce L., P. Eng.

Cooper, Phontroy, P. Eng.
Corbett, Francis M., P. Eng.
Corcoran, Donald J., P. Eng.
Corneau, Raynald, P. Eng.
Costello, James P., P. Eng.
Costello, Anthony B., P. Eng.
Cote, Pierre, P. Eng.
Cote, Stephane, P. Eng.
Coulson, Donald M., P. Eng.
Courage, P. Adam, P. Eng.
Courage, Dale J., P. Eng.
Couves, Brian M., P. Eng.
Cox, John, P. Eng.
Craig, Michelle N., P. Eng.
Cramm, Franklyn M., P. Eng.
Crane, John Charles, P. Eng.
Cranford, Wayne R., P. Eng.
Crawford, Sherwood F., P. Eng.
Crepeau, Louis, P. Eng.
Crewe, Gary Scott, P. Eng.
Crichton, Andrew G., P. Eng.
Crocker, Peter D., P. Eng.
Crosbie, William B., P. Eng.
Crosbie, H. Scott, P. Eng.
Crosbie, Roger M., P. Eng.
Cross, John, P. Eng.
Crossley, Roland V., P. Geo.
Crowther, Kelvin A., P. Eng.
Cumming, Ewan H., P. Geo.
Curran, Terrance, P. Eng.
Curran, Paul, P. Eng.
Curtis, Martin B., P. Eng.
Cutler, I. Marc, P. Geo.
Dale, Terry B., P. Eng.
Daley, Claude G., P. Eng.
Dalley, Donald A., P. Eng.
Daniels, Jason A., P. Eng.
Dasie, William, P. Eng.
Davenport, Peter H., P. Geo.
Davidson, J. Eric, P. Eng.
Davidson, Alan Cecil, P. Eng.
Davies, Stephen, P. Eng.
Davies, David J., P. Eng.
Davies, Michael Huw, P. Eng.
Davis, Robert Frank, P. Eng.
Davis, Robert, P. Eng.
Davis, H. Laban, P. Eng.
Davis, R. Arthur, P. Eng.
Davis, Brendan F., P. Eng.
Davis, James J., P. Eng.
Dawe, Gerald, P. Eng.
Dawe, David W., P. Eng.
Dawe, Anthony A., P. Eng.
Dawe, John G., P. Eng.
Dawe, Lesley A., P. Eng.
Dawe, Byron R., P. Eng.
Dawe, Barry J., P. Eng.
Dawe, C. Roy, P. Eng.
Dawe, Robert D., P. Eng.
Dawson, Kevin J., P. Eng.
Day, John Tariq, P. Eng.
Day, Dale, P. Eng.
Day, Barry C., P. Eng.
Daye, Robert D., P. Eng.
Daye, Bryan J., P. Eng.
De Koos, Peter, P. Eng.

Dean, Mel, P. Eng.
Dean, Kenneth B., P. Eng.
Dearin, Charles, P. Geo.
Dearing, J. Dean, P. Eng.
Debourke, Darren P., P. Eng.
Debruijn, Gunnar G., P. Eng.
Dec, Tomasz, P. Geo.
Decker, H. Shawn, P. Eng.
Decker, Jeremy, P. Eng.
Deering, Paul, P. Eng./P. Geo.
Delaney, Phonse, P. Eng.
Delaney, Paul, P. Geo.
Delaney, Percy, P. Eng.
Demik, James Henry, P. Eng.
Dempster, Robert T., P. Eng.
Denny, Bruce J., P. Eng.
Derks, Barry A., P. Eng.
Desilva, Neil R., P. Geo.
Desjardins, Christian, P. Eng.
Desnoyers, Donald W., P. Geo.
Dessureault, Michel, P. Eng.
Dessureault, Pierre, P. Eng.
Deveau, Marcel L., P. Eng.
Dewitt, Michael, P. Eng.
Dicesare, Donald J., P. Eng.
Dickinson, Geoffrey R.E., P. Eng.
Dickson, W. Lawson, P. Geo.
Didur, Robert Steve, P. Eng.
Dignard, Suelynn E., P. Eng.
Dillabough, Graham D., P. Geo.
Dillon, Randell G., P. Eng.
Dimmell, Norm, P. Eng.
Dimmell, Peter, P. Geo.
Dinn, Gary J., P. Eng.
Dlugosch, Gunter M., P. Eng.
Dominie, Kenneth, P. Eng.
Dormody, Sheri-lynn, P. Eng.
Dormody, Michael J., P. Eng.
Doucet, Serge J., P. Eng.
Doucet, Ronald P., P. Eng.
Doucet, J. Michael, P. Eng.
Douthwright, Trevor, P. Eng.
Dowden, Rodney C., P. Eng.
Downer, Bruce R., P. Eng.
Downey, Cyril J., P. Eng.
Downton, Eric W., P. Eng.
Downton, Neil A., P. Eng.
Doyle, John A., P. Eng.
Doyle, James J., P. Eng.
Drake, Leo, P. Eng.
Driscoll, Donald W., P. Eng.
Driscoll, David W., P. Eng.
Drodge, Ronald M., P. Eng.
Drodge, Gregory, P. Eng.
Drover, Brian, P. Eng.
Drover, Frederick J., P. Eng.
Drover, Kenneth A., P. Eng.
Drover, Geoffrey J.T, P. Eng.
Drover, Lewis Wayne, P. Eng.
Drover, John T., P. Eng.
Drummond, Bryan Walter, P. Eng.
Duffett, Walter Paul, P. Eng.
Duffy, Troy R., P. Eng.
Dufresne, Michel W., P. Geo.
Dufresne, Philippe, P. Eng.
Dugal, Roger, P. Eng.

Duggan, Gerald W., P. Eng.
Duggan, Kenneth F., P. Eng.
Duke, Robert F., P. Eng.
Dumaresq, Peter D., P. Eng.
Dumaresque, Clayton, P. Geo.
Dumont, Gilbert, P. Eng.
Duncan, David, P. Geo.
Duncan, James Gordon, P. Eng.
Dunphy, Gerard M., P. Eng.
Dunphy, Joseph L., P. Eng.
Dunphy, Patricia M., P. Eng.
Dunsworth, Sherry M., P. Geo.
Dupre, Marc, P. Eng.
Durnford, George S., P. Eng.
Dutton, Catherine R., P. Eng.
Dutton, John Joseph, P. Eng.
Dwyer, K. Earl, P. Eng.
Dwyer, Michael J., P. Eng.
Dwyer, B. Lori, P. Geo.
Dyer, Kenneth, P. Eng.
Dyer, Colin J., P. Eng.
Dyer, J. Howard, P. Eng.
Dyke, Stephen D., P. Eng.
Dyke, Sidney G., P. Eng.
Dykeman, Mark R., P. Eng.
Earle, William R.V., P. Eng.
Eastman, E. Dennis, P. Eng.
Eaton, C. Fraser, P. Eng.
Eaton, W. Norris, P. Eng.
Eavis, Charles G., P. Eng.
Eddy, Donald W., P. Eng.
Edmunds, Albert, P. Eng.
Edwards, Paul B., P. Eng.
Edwards, Franklyn, P. Eng.
Edwards, Julian M., P. Eng.
Edwards, Ian G., P. Eng.
Edwards, Craig, P. Eng.
Eichler, George, P. Eng.
Eid, Bassem, P. Eng.
El-Tahan, Hussein, P. Eng.
El-Tahan, Mona S., P. Eng.
El-Gohary, Samir, P. Eng.
El-Mongi, Kadri, P. Eng.
Eldem, H. Tayfun, P. Eng.
Elliott, C. David, P. Eng.
Elliott, Cyril R., P. Eng.
Emberley, Geoffrey C., P. Eng.
Enanny, Faiza, P. Eng.
English, Joseph G., P. Eng.
English, Gerald, P. Geo.
Erzincioglu, Zakaria, P. Eng.
Eschuk, Gregory A., P. Eng.
Evans, John G., P. Eng.
Evans, David, P. Geo.
Evans-Lamswood, Dawn, P. Geo.
Bevely, D. Garth, P. Eng.
Evenson, B. James, P. Eng.
Ewida, Ahmed, P. Eng.
Eyre, Dale G., P. Eng.
Fagan, Alphonsus, P. Geo.
Fahey, John D., P. Eng.
Famery, Michel, P. Eng.
Fancey, Sean P., P. Eng.
Fancey, Everett G., P. Eng.
Farmer, Randy D., P. Geo.
Farrow, Gregory, P. Eng.

Feaver, W. Stephen, P. Eng.
Feehan, Peter, P. Eng.
Fekete, Thomas, P. Eng.
Ferguson, Earle, P. Eng.
Ferry, Peter, P. Eng.
Finch, Donald R., P. Eng.
Finelli, Donato, P. Eng.
Finn, William R., P. Eng.
Fisher, Andrew, P. Eng.
Fitzgerald, John G., P. Eng.
Fitzpatrick, Donald, P. Geo.
Fitzpatrick, Denis S., P. Geo.
Fleming, William, P. Eng.
Fleming, Gerald J., P. Eng.
Fleming, John M., P. Geo.
Fletcher, David Edward, P. Eng.
Fleurant, Mario, P. Eng.
Flood, Roger D., P. Eng.
Flynn, Craig G., P. Eng.
Flynn, John, P. Eng.
Flynn, P. Robert, P. Eng.
Fogwill, David W., P. Eng.
Fogwill, Lawrence, P. Eng.
Follett, Douglas R., P. Eng.
Follett, Gary J., P. Eng.
Fong, David J., P. Eng.
Fontaine, Yves Jean, P. Eng.
Foote, Wesley, P. Eng.
Forbes, Robert G., P. Eng.
Forbes, Donald S., P. Eng.
Forbes, Cyril R., P. Eng.
Forbrigger, Blair F., P. Eng.
Forsey, Paul K., P. Eng.
Forsey, Reginald B., P. Eng.
Forward, Colin Atwood, P. Eng.
Foster, David W., P. Eng.
Foster, Keith S., P. Eng.
Fowlow, D. Chad, P. Eng.
Fox, David John, P. Eng.
Fradsham, Ian, P. Eng.
Franey, Ronald P., P. Eng.
Freake, Howard, P. Eng.
Freeman, S. John, P. Geo.
Freeman, Geoffrey W., P. Eng.
French, H. Graham, P. Eng.
French, Victor A., P. Geo.
French, Derrick, P. Eng.
French, James K., P. Eng.
Freud, Michael, P. Eng.
Frew, Allan, P. Geo.
Friis, Dag A., P. Eng.
Froude, Timothy, P. Geo.
Fry, Wallace W., P. Eng.
Fry, Jeffrey, P. Eng.
Fudge, Lloyd D., P. Eng.
Furlong, Christopher, P. Eng.
Furst, Jan, P. Eng.
Fyfe, Lorie, P. Eng.
Gagnon, Michel, P. Eng.
Gaherty, W. Denis, P. Eng.
Gale, John E., P. Eng./p. Geo.
Galvez, Cesar A., P. Eng.
Gardiner, Terence J., P. Eng.
Gates, Andrew, P. Eng.
Gaudet, Paul E., P. Eng.
Gauthier, Peter Leo, P. Eng.

Gelman, Yury, P. Eng.
Gemmell, Donald E., P. Geo.
George, F. Newman, P. Eng.
George, Arthur L., P. Eng.
Ghazala, Sue, P. Eng.
Ghica, Viorel, P. Eng.
Giangrande, Aldo G., P. Eng.
Gibbons, Rex, P. Geo.
Gibling, Glen, P. Eng.
Giffin, James Murray, P. Eng.
Gill, Parmbir S., P. Eng.
Gill, Robert J., P. Eng.
Gill, David G., P. Eng.
Gillespie, Helen, P. Geo.
Gillespie, Randall T., P. Geo.
Gillespie, Corinne M., P. Eng.
Gillespie, Frank J., P. Eng.
Gillis, Nicholas, P. Eng.
Gillis, Michael, P. Eng.
Ginn, Kevin E., P. Eng.
Girard, Pierre, P. Eng.
Glavine, Michael G., P. Eng.
Goddard, Grant, P. Eng.
Goebel, Martin G., P. Eng.
Gomez, Emilio E., P. Eng.
Good, Donald C., P. Eng.
Goodland-Hennessey, Valerie, P. Eng.
Goodman, Paul R., P. Eng.
Goodridge, Douglas N., P. Eng.
Goodyear, Clifford A., P. Eng.
Goodyear, Terence S., P. Eng.
Goosney, David D., P. Eng.
Goosney, John E., P. Eng.
Goosney, Richard F., P. Eng.
Goosney, Jennifer Ann, P. Eng.
Goosney, Robert J., P. Eng.
Gordon, James L., P. Eng.
Gorman, Joseph, P. Eng.
Gorman, Michael C., P. Eng.
Gorman, J. Clifford, P. Eng.
Gosine, Raymond G., P. Eng.
Gosine, Raymond F., P. Eng.
Gosse, R. Gordon, P. Eng.
Gosse, Gary R., P. Eng.
Gosse, Robert M., P. Eng.
Gosse, Michelle M., P. Eng.
Gosse, K. Richard, P. Eng.
Gosse, Wayne H., P. Eng.
Goudie, Herbert W., P. Eng.
Goulding, Kevin, P. Eng.
Goulding, William J., P. Eng.
Gover, Darrell C., P. Eng.
Gover, Douglas J., P. Eng.
Gover, Paul, P. Eng.
Grabke, Hermann K., P. Eng.
Grabowski, Ludwik, P. Eng.
Graham, Philip, P. Eng.
Grainger, Andrew J.e., P. Eng.
Grainger, Stephen P., P. Eng.
Grandy, William, P. Eng.
Grandy, Lisa D., P. Eng.
Grant, Hugh M., P. Eng.
Grant, Frank D., P. Eng.
Granter, Clyde C., P. Eng.
Gravel, J.R. Douglas, P. Eng.
Graves, Garth D., P. Geo.

Graves, R. Mark, P. Geo.
Gray, Eric, P. Eng.
Gray, Bruce, P. Eng.
Greeley, Aubrey W., P. Eng.
Green, Karl W., P. Eng.
Green, Jeffrey D., P. Eng.
Green, Allan C., P. Eng.
Green, James F., P. Eng.
Greene, A. Bradley, P. Eng.
Greening, Dawna, P. Eng.
Greenland, George, P. Eng.
Greer, George J., P. Eng.
Gregory, Martin, P. Eng.
Grenier, Stephane, P. Eng.
Griffin, Paul J., P. Eng.
Griffin, William C., P. Eng.
Grossert, Alan J., P. Eng.
Groves, Glen C., P. Eng.
Guigne, Jacques Y., P. Geo.
Guiney, Joseph, P. Eng.
Guo, Ping, P. Eng.
Guzzwell, Keith, P. Geo.
Guzzwell, John A., P. Eng.
Ha, Joseph T.K., P. Eng.
Haddara, Mahmoud, P. Eng.
Halabieh, Bassam, P. Eng.
Haldar, Asim Kumar, P. Eng.
Hall, John Charles, P. Eng.
Hall, Dennis, P. Eng.
Hall, Jeremy, P. Geo.
Hall, Francis W., P. Eng.
Hallett, Frederick W, P. Eng.
Hamilton, Paul R., P. Eng.
Hamilton, L. Wayne, P. Eng.
Hammerschlag, Mark, P. Eng.
Hammond, Karl W., P. Eng.
Hancock, F. Scott, P. Eng.
Hancock, Tony A., P. Eng.
Hancock, Krista, P. Eng.
Hancock, Steven H., P. Eng.
Hann, Karl, P. Eng.
Hann, Lewis G., P. Eng.
Hannah, Arthur W., P. Eng.
Hannon, Keith H., P. Eng.
Hanrahan, Randy G., P. Eng.
Harbin, Andrew C.H., P. Eng.
Harder, John Allan, P. Eng.
Hardie, Andrew P., P. Eng.
Harnois, Patrice, P. Eng.
Harnum, David J., P. Eng.
Haroun, Ibrahim A., P. Eng.
Harris, James, P. Geo.
Harris, David, P. Eng.
Harris, Alvin, P. Geo.
Harris, Carl J.J., P. Eng.
Harris, Anthony C., P. Eng.
Harris, Thomas J., P. Eng.
Harrison, Iain Michael, P. Eng.
Harrison, Donald C., P. Eng.
Harriss, Neville D., P. Eng.
Hartley, Dennis, P. Eng.
Harty, James J., P. Eng.
Harvey, Robert M., P. Eng.
Harvey, Robert H., P. Eng.
Hatcher, C. Paul, P. Eng.
Hatt, P. Kenneth, P. Eng.

Hattie, Donald W., P. Geo.
Hawboldt, Kelly Anne, P. Eng.
Hawco, Gerald V., P. Eng.
Hawkins, David, P. Geo.
Hawkins, Duane Paul, P. Eng.
Hayes, Jeffrey C., P. Eng.
Hayes, John, P. Geo.
Hayes, Alvin, P. Eng.
Haynes, James R., P. Eng.
Haynes, Donald J., P. Eng.
Haynes, Wynyard, P. Eng.
Hayward, Richard, P. Eng.
Hayward, Kenneth, P. Eng.
Healy, Michael L., P. Eng.
Hearn, Paul M., P. Eng.
Hebb, Stephen, P. Eng.
Hedges, Reginald, P. Eng.
Heffernan, Mary J., P. Eng.
Helwig, Philip C., P. Eng.
Henderson, Lorne C., P. Eng.
Henderson, Robert John, P. Eng.
Henderson, John P., P. Eng.
Henley, John J., P. Eng.
Hennebury, Peter F., P. Eng.
Henney, Basil Scot, P. Eng.
Henry, Charles W., P. Eng.
Hermanski, Grzegorz, P. Eng.
Hermanto, Ivi, P. Eng.
Hewitt, Clarence W., P. Eng.
Hewlett, David G., P. Eng.
Hibbs, Richard, P. Eng.
Hickey, Michael D., P. Eng.
Hickman, Keith W., P. Eng.
Hicks, David, P. Eng.
Hicks, Holly, P. Eng.
Hicks, Larry, P. Geo.
Hicks, Frederick T., P. Eng.
Higdon, Stephen, P. Eng.
Hildebrand, Ronald G., P. Eng.
Hill, David J., P. Eng.
Hillier, K. Barry, P. Eng.
Hillier, Christine M., P. Eng.
Hillier, William G., P. Eng.
Hinchey, Danu W.f., P. Eng.
Hinchey, Michael J., P. Eng.
Hinchey, Morgan F., P. Eng.
Hinchey, James P., P. Geo.
Hirou, Martine, P. Eng.
Hiscock, Stephen, P. Eng.
Hiscock, Keith N., P. Eng.
Hiscock, Elizabeth, P. Eng.
Hiscock, D.E. Todd, P. Eng.
Hiscock, Barry C., P. Eng.
Hiscott, Richard, P. Geo.
Hoadley, Gary Robert, P. Eng.
Hobbs, Geoffrey, P. Eng.
Hodder, David E., P. Eng.
Hodder, Jeffrey W., P. Eng.
Hodder, Charles M., P. Eng.
Hoddinott, W. Paul, P. Eng.
Hoddinott, Terry K., P. Eng.
Hodge, Steven Eric, P. Eng.
Hodgson, Gurney J., P. Geo.
Hogan, Michael J., P. Eng.
Hogan, Jane K., P. Eng.
Hogan, Thomas G., P. Eng.

Hogan, Noel E.k., P. Eng.
Holden, Glenn Dean, P. Eng.
Holden, Gordon J., P. Eng.
Hollett, Phillip, P. Eng.
Hollett, Timothy, P. Eng.
Holm, Carl, P. Eng.
Holmes, Jeffrey D., P. Eng.
Homer, John, P. Eng.
Honarvar, Pauline, P. Geo.
Honeygold, Peter, P. Eng.
Hong, Derrick, P. Eng.
Honsberger, Douglas K., P. Eng.
Hood, David, P. Eng.
Hood, Charles H., P. Eng.
Hookey, Neil A., P. Eng.
Hopkins, Robert M., P. Eng.
Hopkins, G. Duane, P. Eng.
Hopkins, Michael T., P. Eng.
Opkins, Robert M., P. Eng.
House, Glenn L., P. Geo.
House, Marlowe, P. Eng.
Houze, Graham C., P. Eng.
Howell, Dwight G., P. Eng.
Howell, Gordon E., P. Eng.
Howell, G. Boyd, P. Eng.
Howley, Michael F., P. Eng.
Hruska, Vlado, P. Eng.
Hryniw, William J., P. Eng.
Huang, Ted T., P. Eng.
Hubeny, Jiri K., P. Eng.
Hubley, Fred G.A., P. Eng.
Hudson, John F., P. Eng.
Huffman, Robert A., P. Eng.
Hughes, B. Yvette, P. Eng.
Hull, Greg, P. Eng.
Humber, David F., P. Eng.
Humber, Gerard, P. Eng.
Humby, Gary K., P. Eng.
Humphries, Paul W., P. Eng.
Hunt, Jeffrey P., P. Geo.
Hunt, Augustus R., P. Eng.
Hunt, William C., P. Eng.
Hunt, Kevin A., P. Eng.
Hunt, Douglas C., P. Eng.
Hunt, Neil C., P. Eng.
Hunter, Montgomery C, P. Eng.
Hupman, Murray, P. Eng.
Hurley, David M., P. Eng.
Hurley, Shawn, P. Eng.
Hurley, Suzanne, P. Eng.
Husain, Tahir, P. Eng.
Hussein, Ismail, P. Eng.
Hussein, Amgad Ahmed, P. Eng.
Hussey, David F., P. Eng.
Hutchens, Donald L., P. Eng.
Hutchings, Christopher, P. Geo.
Hutton, Peter A., P. Eng.
Huxter, Frank J., P. Eng.
Hyde, Darryl G., P. Eng.
Hynes, J.A. Francis, P. Eng.
Hynes, William G., P. Eng.
Hynes, Todd P., P. Eng.
Hynes, Todd G., P. Eng.
Hynes, Garret J., P. Eng.
Hynes, Daniel G., P. Eng.
Hynes, Gordon J., P. Eng.

Hynes, Ivan C., P. Eng.
Hynes, Gary M., P. Eng.
Iddamsetty, Krishna, P. Eng.
Imhoff, Barry A., P. Eng.
Inkpen, Stuart L., P. Eng.
Ireland, Hughie A., P. Eng.
Isenor, Randal, P. Eng.
Isenor, Gerald, P. Eng.
Jackson, Roger S., P. Geo.
Jacobs, Bradford, P. Eng.
Jacobs, Wallace K., P. Eng.
Jagodits, Francis L., P. Eng./P. Geo.
James, Dr. Donald, P. Geo.
Janes, Colin W., P. Eng.
Janes, Michael B., P. Eng.
Jani, Rajendra, P. Eng.
Jansen, Hendrik, P. Eng.
Jardine, Michael R., P. Eng.
Jeans, Philip K., P. Eng.
Jeans, David G., P. Eng.
Jeffrey, Norman, P. Eng.
Jenkins, Jeffrey G., P. Eng.
Jenkins, Nolan, P. Eng.
Jenkins, Robert H.D., P. Eng.
Jenner, George A., P. Geo.
Jerome, Pierre, P. Eng.
Jerrett, Robert C., P. Eng.
Jerrett, Eric K., P. Eng.
Jewer, Kenneth H.G., P. Eng.
Jewer, Robert W., P. Eng.
Jeyasurya, Benjamin, P. Eng.
Jiang, Dajiu, P. Eng.
Jin, Gordon, P. Eng.
Johnson, Ronald S., P. Eng.
Johnson, Maxwell, P. Eng.
Johnston, Donald R., P. Eng.
Johnstone, Terry J., P. Eng.
Jonasson, William B., P. Eng.
Jones, Gordon, P. Eng.
Jones, Brent M., P. Eng.
Jones, Russell L., P. Eng.
Jones, Gregory L., P. Eng.
Jones, Richard, P. Eng.
Jordaan, Ian J., P. Eng.
Jost, Roland M., P. Eng.
Journeaux, Noel L., P. Eng./P. Geo.
Julien, Michel, P. Eng.
Kaderali, Ayiaz, P. Geo.
Kalsi, Harsimran, P. Eng.
Kantarcioglu, M. Vehbi, P. Eng.
Karasek, Colin J., P. Eng.
Karyampudi, Rayalu, P. Eng.
Kaushik, Raj K., P. Eng.
Kavanagh, Dion James, P. Eng.
Kavanagh, Sean J., P. Eng.
Kean, Jason R., P. Eng.
Kean, Baxter, P. Geo.
Kearley, Robert, P. Eng.
Keating, Kevin P., P. Eng.
Keating, Keith A., P. Eng.
Keating, Kimberly Ann, P. Eng.
Keating, James P., P. Eng.
Keats, Corey G., P. Eng.
Keats, Willie, P. Eng.
Keefe, C. Michael, P. Eng.
Keeping, Lionel M., P. Eng.

Kehler, Gerald Ian, P. Eng.
Kellestine, Wilfred M., P. Eng.
Kelly, Gerard C., P. Eng.
Kelly, David, P. Eng.
Kelly, Robert J., P. Eng.
Kelly, Brian E., P. Eng.
Kelly, Robert, P. Geo.
Kelly, G. Francis, P. Eng.
Kelsey, Jane, P. Eng.
Kemp, David, P. Eng.
Kemp, Douglas Roy, P. Eng.
Kendall, H. Thomas, P. Eng.
Kendall, Keith L., P. Eng.
Kennedy, Edward W., P. Eng.
Kennedy, John M., P. Eng.
Kennedy, Robert J., P. Eng.
Kennedy, Gary, P. Eng.
Kennedy, James G., P. Eng.
Kennedy, William J., P. Eng.
Kennedy, Robin M., P. Eng.
Kennedy, William F., P. Eng.
Kenny, Robert T., P. Eng.
Kepka, Jan, P. Eng.
Kerr, Ian Ross, P. Eng.
Kerr, Andrew, P. Geo.
Khan, Hasin U., P. Eng.
Kieley, Robert J., P. Eng.
Kieley, Jane M., P. Eng.
Kieley, John, P. Geo.
Kierans, Thomas W., P. Eng.
Kiernan, Aidan E., P. Eng.
Kiley, Liam, P. Eng.
Kim, H. Joe, P. Eng.
King, Arthur P., P. Geo.
King, Danny, P. Eng.
King, Mark S., P. Geo.
King, Carl Boyd, P. Eng.
King, Jeffery C., P. Eng.
King, Geoffrey R., P. Eng.
King, Gene R., P. Eng.
King, Terry P., P. Eng.
King, Robin W., P. Eng.
King, Stephen H., P. Eng.
King, Anthony, P. Eng.
Kingsley, Reginald G., P. Eng.
Kirby, Christopher, P. Eng.
Kirby, J. Alan, P. Eng.
Kirby, Frederick, P. Geo.
Knee, James, P. Eng.
Knight, Kenneth, P. Eng.
Knight, J. Carl, P. Eng.
Knox, Leonard P., P. Eng.
Kohut, Jerry, P. Eng.
Koniecki, Mariusz, P. Eng.
Kowalski, Wlodzimierz, P. Eng.
Krein, Heinz L., P. Eng.
Krickler, Claudio, P. Eng.
Kunchur, Gopal R., P. Eng.
Kuta, Robert M. W., P. Eng.
Kwan, Gordon, P. Eng.
La Cour, K. Sean, P. Eng.
Lacasse, Francis, P. Eng.
Lacey, Michael P., P. Eng.
Lachance, Claire, P. Eng.
Lacoursiere, Jean, P. Eng.
Lacroix, Kevin Joseph, P. Eng.

Lahey, Paul J., P. Eng.
Lahey, Edmund, P. Eng.
Laidley, Thomas, P. Geo.
Lalancette, Carl, P. Eng.
Lalonde, Richard D., P. Eng.
Lambert, Dino R., P. Eng.
Lammey, John, P. Eng.
Landra, Thomas Keith, P. Eng.
Landry, Joel, P. Eng.
Landva, Jorn, P. Eng.
Lane, Thomas, P. Geo.
Langdon, Shawn, P. Eng.
Langins, Ernests A., P. Eng.
Langley, Wilbert S., P. Eng.
Langridge, Robert John, P. Geo.
Laracy, Patrick J., P. Geo.
Lata, Edward H., P. Eng.
Lau, Alex Y-H., P. Eng.
Laufers, Juris R., P. Eng.
Laurentius, Theodore B.A, P. Eng.
Lawless, Perry Bruce, P. Eng.
Layden, Albert G., P. Eng.
Lebeau, Marcel, P. Eng.
Leblanc, Darin Joseph, P. Eng.
Ledrew, Terry P., P. Eng.
Ledrew, Jeffrey R., P. Eng.
Ledrew, Ronald H., P. Eng.
Lee, Joseph James, P. Eng.
Lee, Jason R., P. Eng.
Lee, Daniel V., P. Geo.
Leeman, Robert L., P. Eng.
Lefeuvre, E. Patricia, P. Eng.
Legge, Neil L., P. Eng.
Legge, Richard V., P. Eng.
Leggo, Richard W., P. Eng.
Leja, Gunar, P. Eng.
Lemay, Iain, P. Eng.
Lemessurier, Brian S., P. Eng.
Lemessurier, Peter J., P. Eng.
Leonard, Keith A., P. Eng.
Leonard, John J., P. Eng.
Leriche, Timothy, P. Eng.
Lester, Peter D., P. Eng.
Lethbridge, Steven A., P. Eng.
Leung, Arthur Ki-ki, P. Eng.
Levasseur, Michel, P. Eng.
Lever, Gregory, P. Eng.
Lewis, David Victor, P. Eng.
Lewis, Jerome T., P. Eng.
Lewis, David J., P. Eng.
Li, Karl, P. Eng.
Liebich, Tomas, P. Eng.
Lin, Frank Fang-K, P. Eng.
Linehan, Valentine J., P. Eng.
Linfield, Dana C., P. Eng.
Little, Chesley L., P. Eng.
Little, Gregory, P. Eng.
Liu, Dr. Pengfei, P. Eng.
Liverman, David, P. Geo.
Livet, Robert G., P. Eng.
Locke, Stephen, P. Eng.
Lockyer, David George, P. Eng.
Loder, Penny Gail, P. Eng.
Lodge, Stephen A., P. Eng.
Lomond, Brian V., P. Eng.
Lone, Kato, P. Eng.

Long, James D., P. Eng.
Lorenzen, William A., P. Eng.
Lu, Zhong Qun, P. Eng.
Ludlow, Earl, P. Eng.
Lue Choy, Joseph, P. Eng.
Luffman, Brian, P. Eng.
Lundrigan, George W., P. Eng.
Lundrigan, Harold W., P. Eng.
Lush, Philip L., P. Eng.
Luznik, Luksa, P. Eng.
Lye, Leonard, P. Eng.
MacDonald, Michael S., P. Eng.
MacDonald, Harry F., P. Eng.
MacDonald, Michael R., P. Eng.
MacDonald, Donald Ewen, P. Eng.
MacDonald, Andrew, P. Eng.
MacDonald, Mario R., P. Eng.
MacDonald, M. Daniel, P. Eng.
MacDonald, Brandon J., P. Eng.
MacDonald, Kenna, P. Eng.
MacDonald, Joseph Alan, P. Eng.
MacDonald, J.A. Gordon, P. Eng.
MacDougall, Craig S., P. Geo.
MacGillivary, Gary F., P. Eng.
MacGillivray, Guy C., P. Geo.
MacGregor, Ian K., P. Eng.
MacIsaac, Robert, P. Eng.
MacIsaac, W. Bernard, P. Geo.
MacKenzie, Richard I.R, P. Eng.
MacKey, Anthony L., P. Eng.
MacKey, Paul T., P. Eng.
MacLaggan, Lawrence, P. Eng.
MacLean, Alexander J., P. Eng.
MacLean, Charles R., P. Eng.
MacLeod, Robert I.W., P. Geo.
MacLeod, Ian B., P. Eng.
MacNeill, Andrew E., P. Eng.
MacPherson, William A., P. Eng.
MacVicar, Kenneth B., P. Eng.
Maddigan, Patrick J., P. Eng.
Maddocks, Derrick K., P. Eng.
Mader, Thomas E., P. Eng.
Madryga, Alexander, P. Eng.
Maguire, Michael J., P. Eng.
Mahon, Patrick J., P. Eng.
Mahon, James W., P. Eng.
Majchrowski, Barbara K., P. Eng.
Mak, Lawrence M., P. Eng.
Mallam, John E., P. Eng.
Mallam, Carl E., P. Eng.
Maloney, Wilfred G., P. Eng.
Maloney, John, P. Eng.
Maloney, Douglas, P. Eng.
Maloney, Wilfred P., P. Eng.
Manning, David W., P. Eng.
Manocha, Jug S., P. Eng.
Mantey, Martin J., P. Eng.
Manuel, Jacqueline, P. Eng.
Manuel, Wayne G., P. Eng.
March, Brian J., P. Eng.
March, Roger B., P. Geo.
March, Roger P., P. Eng.
Marche, Alberta M., P. Eng.
Marche, Sheldon, P. Eng.
Marks, Larry, P. Eng.
Marks, Gerald G., P. Eng.

Marshall, Christopher, P. Eng.
Marshall, H. Stanley, P. Eng.
Marshall, Mervin A., P. Eng.
Marshall, Alfred R., P. Eng.
Martin, John F., P. Eng.
Martin, Frederick H., P. Eng.
Martin, A. Denis, P. Eng.
Martin, David C., P. Eng.
Martin, Willis P., P. Eng.
Martinovic, Tom, P. Eng.
Marzouk, Hesham M.E., P. Eng.
Mason, Roger, P. Geo.
Mastroberardino, Carlo, P. Eng.
Matchem, Jerry, P. Eng.
Matchim, Lester R., P. Eng.
Matich, Miroslav, P. Eng.
Matthews, Donald G., P. Eng.
Matthews, Raymond G., P. Eng.
Matthews, Kenneth J., P. Eng.
Maximchuk, Alan Jacob, P. Eng.
May, Charles W., P. Eng.
Maybee, William, P. Eng.
Mayberry, William R., P. Eng.
Mazour, Alexander, P. Eng.
Mazza, Ronald, P. Eng.
McArthur, Elaine, P. Eng.
McArthur, Brian A., P. Eng.
McCarthy, Cyril J., P. Eng.
McCarthy, Terrence P., P. Eng.
McCarthy, Shaun L., P. Eng.
McCavour, Scot S., P. Eng.
McCharles, Grant G., P. Eng.
McClintock, Alfred, P. Eng.
McConnell, Douglas L., P. Eng.
McDonald, Paul J.P., P. Eng.
McDonald, Randy D., P. Eng.
McDougall, Alvin Ray, P. Eng.
McEwen, William D., P. Eng.
McGerrigle, James R., P. Eng.
McGowan, Brian W., P. Eng.
McGrath, Bruce, P. Eng.
McGrath, John W., P. Eng.
McGuire, Robert, P. Eng.
McHattie, Leslie D., P. Eng.
McIntyre, Judith, P. Geo.
McKenna, Richard F., P. Eng.
McKenney, Gregory, P. Eng.
McKenzie, Colin, P. Geo.
McLean, Robert Clyde, P. Eng.
McLean, Jill, P. Geo.
McLean, Stephen M., P. Eng.
McLure, A. Robert, P. Eng.
McNeill, Stanley, P. Eng.
McNiven, Blair J., P. Eng.
McPherson, Daniel, P. Eng.
McTavish, Stuart, P. Eng.
Meadus, Harry M., P. Eng.
Meaney, Richard B., P. Eng.
Meaney, Gerard R., P. Eng.
Medcalf, James Thomas, P. Eng.
Mehta, Angelina, P. Eng.
Meisen, Axel, P. Eng.
Melendy, William G., P. Eng.
Mensch, William, P. Eng.
Mercer, Cluney G., P. Eng.
Mercer, Robert C.J., P. Eng.

Mercer, Sandra, P. Eng.
Mercer, Scott, P. Eng.
Mercer, Ron C., P. Eng.
Mercer, Todd Calvin, P. Eng.
Mercer, Roderick, P. Geo.
Mercer, Maxwell, P. Eng.
Mercer, Kimberly A., P. Eng.
Mercer, Eric W., P. Eng.
Meunier, Denis, P. Eng.
Mews, Christopher, P. Eng.
Meyer, James, P. Geo.
Miazga, Greg W., P. Eng.
Miles, Calvin, P. Geo.
Miles, Harold N., P. Eng.
Miles, Darrell J., P. Eng.
Milke, Fred L., P. Eng.
Millan, James P., P. Eng.
Millan, David E.L., P. Eng.
Millan, Steven, P. Geo.
Miller, Raymond R., P. Eng.
Miller, Hugh G., P. Geo.
Millette, Denis F. J., P. Eng.
Mills, Walter F., P. Eng.
Mills, Chesley, P. Eng.
Mills, William P., P. Eng.
Mills, Darryl, P. Eng.
Mills, Archie L., P. Eng.
Milne, William J., P. Eng.
Mistry, Bhana D., P. Eng.
Mitchell, David R., P. Eng.
Mitchell, David W., P. Eng.
Mitchelmore, Perry, P. Eng./P. Geo.
Molgaard, Johannes, P. Eng.
Molloy, John F., P. Eng.
Molloy, Paul W.P., P. Eng.
Monahan, Craig, P. Eng.
Moncur, Mark C., P. Eng.
Montague, H. Edmund, P. Geo.
Moody, Douglas B., P. Eng.
Mooney, Karl J., P. Eng.
Moore, Darren, P. Eng.
Moore, Fraser L., P. Eng.
Moore, George S., P. Eng.
Moore, Edward, P. Eng.
Moore, Paul Jeffrey, P. Geo.
Moores, Stanley E., P. Eng.
Moores, Weldon G., P. Eng.
Moores, R. David, P. Eng.
Moores, Laurence G., P. Eng.
Moores, Gerald W., P. Eng.
Morin, Gaetan, P. Eng.
Morin, Julien, P. Eng.
Morris, H. Mervin, P. Eng.
Morris, Denis A., P. Eng.
Morris, Douglas R.M., P. Eng.
Morrison, Stephen L., P. Eng.
Morrison, Timothy D., P. Eng.
Morrison, Luke Stewart, P. Eng.
Morrison, Kenneth I., P. Eng.
Morrison, Stuart Allan, P. Eng.
Morrissey, John P., P. Eng.
Moscovitch, Daniel, P. Eng.
Mosher, David, P. Eng.
Moss, Gregory, P. Eng.
Moss, Derrick L., P. Eng.
Motty, Stephen, P. Eng.

Mouland, Kevin, P. Eng.
Moulton, Robert J.H., P. Eng.
Mozaffari, Said, P. Eng.
Mugford, Ralph W., P. Eng.
Muggeridge, Karen, P. Eng.
Muir, Robert, P. Eng.
Muise, Stephanie A., P. Eng.
Muise, Jason L., P. Eng.
Mulcahy, John J., P. Eng.
Munaswamy, Katna, P. Eng.
Murley, S. Roy, P. Eng.
Murphy, Kevin John, P. Eng.
Murphy, David E., P. Eng.
Murphy, Noel, P. Eng.
Murphy, Richard T., P. Eng.
Murphy, Maurice J., P. Eng.
Murphy, David E., P. Eng.
Murphy, Robert, P. Eng.
Murphy, T. Clair, P. Eng.
Murray, Gary L., P. Eng.
Murray, John J., P. Eng.
Murrin, Felix J., P. Eng.
Myers, Derrick R., P. Eng.
Myers, Gordon A. Jr, P. Eng.
Myers, David R., P. Eng.
Myers, Roy W., P. Eng.
Myles, Daniel J. C., P. Eng.
Nappert, Daniel, P. Eng.
Neary, George N., P. Eng.
Neil, Danny R. G., P. Eng.
Nell, Peter G., P. Eng.
Nemec, Anthony O., P. Eng.
Nevens, Martin, P. Eng.
Neville, R. John, P. Eng.
Neville, David, P. Eng.
Newbury, David, P. Eng.
Newbury, A. Douglas, P. Eng.
Newbury, Frank E., P. Eng.
Newhook, Robert C., P. Eng.
Newhook, Dennis, P. Eng.
Newhook, Lawrence M., P. Eng.
Newhook, Ronald V., P. Eng.
Newman, Derek, P. Geo.
Newman, William P., P. Eng.
Newton, William J., P. Eng.
Ngui, Phillip, P. Eng.
Nguyen, Van Hiep, P. Eng.
Nicholas, James E., P. Eng.
Nicholson, Kent D., P. Eng.
Nickerson, Bruce A., P. Eng.
Nickerson, Donald R., P. Eng.
Nicoll, Murray, P. Eng.
Nippard, David W., P. Eng.
Nippard, Franklin E., P. Eng.
Nixon, William, P. Eng.
Noah, Gregory, P. Geo.
Noah, Ronald J., P. Eng.
Noel, Chester F., P. Eng.
Noel, Keith A., P. Eng.
Noel, Alexander C., P. Eng.
Noftall, Edward A.G, P. Eng.
Nolan, Paul, P. Eng.
Nolan, Chris D., P. Eng.
Nolan, Francis J., P. Eng.
Norberg, William R., P. Eng.
Noseworthy, Steven D., P. Eng.

Noseworthy, Rick A., P. Eng.
Noseworthy, David W., P. Eng.
Noseworthy, Donald E., P. Eng.
Noseworthy, Charles, P. Eng.
Noseworthy, H. William, P. Eng.
Noseworthy, Robert A., P. Eng.
Nugent, William J., P. Eng.
Nugent, John, P. Eng.
Nwoke, John C., P. Eng.
O'Brien, Kevin G., P. Eng.
O'Brien, Janice L., P. Eng.
O'Brien, Paul J., P. Eng.
O'Brien, Kenneth Ross, P. Eng.
O'Brien, Michael John, P. Eng.
O'Brien, Felicity, P. Geo.
O'Brien, Sean, P. Geo.
O'Brien, Brian, P. Geo.
O'Brien, Michael T., P. Eng.
O'Connell, Brendan, P. Eng.
O'Connell, John G., P. Geo.
O'Dea, Frank P., P. Eng.
O'Grady, Anthony C., P. Eng.
O'Keefe, Edward, P. Eng.
O'Keefe, Edward F., P. Eng.
O'Keefe, Colleen, P. Eng.
O'Keefe, Robert C., P. Eng.
O'Keefe, William A., P. Eng.
O'Keefe, Glenn P., P. Eng.
O'Keefe, Frederick J., P. Eng.
O'Leary, Paul F., P. Eng.
O'Leary, Timothy, P. Eng.
O'Neill, R. Keith, P. Eng.
O'Reilly, Michael G., P. Eng.
O'Reilly, John F., P. Eng.
O'Reilly, Albert E., P. Eng.
O'Rourke, Thomas H., P. Eng.
Oke, Alexander R., P. Eng.
Oldford, Todd Albert, P. Eng.
Osmond, Todd C., P. Eng.
Osmond, W. Robert, P. Eng.
Osmond, J. Dean, P. Eng.
Osmond, Donald P., P. Eng.
Outerbridge, Mark W., P. Eng.
Outerbridge, Peter N., P. Eng.
Palmer, John Hiram, P. Eng.
Panu, Umed S., P. Eng.
Paolini, Neil A., P. Eng.
Paradis, Rejean, P. Eng.
Pardy, N.V. Bruce, P. Eng.
Pardy, John W., P. Eng.
Parent, Pierre M., P. Eng.
Parke, John William, P. Eng.
Parsons, Norman C., P. Eng.
Parsons, Jennifer, P. Eng.
Parsons, Edward, P. Eng.
Parsons, David B., P. Eng.
Parsons, Roy A., P. Eng.
Parsons, William Roy, P. Eng.
Parsons, Rex T., P. Eng.
Pasiecznik, Eugene, P. Eng.
Patel, Rashmi V., P. Eng.
Patel, Avakash, P. Eng.
Paterson, Robert B., P. Eng.
Patton, Douglas G., P. Eng.
Paul, Eric G., P. Eng.
Paul, Ronald, P. Eng.

Paulin, Michael J., P. Eng.
Pauls, Arthur, P. Eng.
Payne, Robert, P. Eng.
Pazos, Miguel Angel, P. Eng.
Peach, Albert, P. Eng.
Peach, Donald E., P. Eng.
Pear, Michael, P. Eng.
Pear, Winston E.N., P. Eng.
Pearcey, Derek E., P. Eng.
Pearson, M. Paul, P. Eng.
Pearson, Robert M., P. Eng.
Pearson, Samuel J., P. Eng.
Peck, Kenneth W., P. Eng.
Peddle, Mark S., P. Eng.
Peddle, Randal K., P. Eng.
Peddle, Kirk, P. Eng.
Peddle, David H., P. Eng.
Peddle, Terence C., P. Eng.
Pedram, Mahyar, P. Eng.
Pelley, Bradley H., P. Eng.
Pelley, Darren, P. Eng.
Penney, Glenn, P. Eng.
Penney, Frederick, P. Eng.
Penney, Douglas, P. Eng.
Penney, Craig S., P. Eng.
Peros, Martin M., P. Eng.
Perry, Ian A., P. Geo.
Peters, David, P. Eng.
Peters, Dennis Keith, P. Eng.
Peters, G. Ross, P. Eng.
Pheeney, Peter, P. Eng.
Phillips, Ryan, P. Eng.
Philpott, Daniel, P. Eng.
Philpott, Paul J., P. Geo.
Philpott, Donald M., P. Eng.
Philpott, J. Myles, P. Eng.
Piatti, Richard, P. Eng.
Picco, Michael, P. Eng.
Picco, Robert C., P. Eng.
Pickett, J. Wayne, P. Geo.
Pickford, Glenn E., P. Eng.
Piercey, Martin L., P. Eng.
Piercy, E. Gerard, P. Eng.
Pieroway, Roy S., P. Eng.
Pike, Christopher, P. Geo.
Pike, Keith L., P. Eng.
Pike, Howard L., P. Eng.
Pike, Kenneth R., P. Eng.
Pike, Gordon J., P. Eng.
Pike, Daniel W., P. Eng.
Pike, Herbert, P. Eng.
Pike, John P., P. Eng.
Pilgrim, Larry, P. Geo.
Pinhorn, William W., P. Eng.
Pinsent, Cyril Mike, P. Eng.
Pippy, Mark, P. Eng.
Piskorski, Andrzej, P. Eng.
Pitcher, Ronald, P. Eng.
Pitt, John K.J., P. Eng.
Pittman, Scott, P. Eng.
Plamondon, Jean-Pierre, P. Eng.
Ploughman, James, P. Eng.
Pollard, Jerry D., P. Geo.
Pollett, R. Troy, P. Eng.
Pond, Bruce, P. Eng.
Poole, Garrett E., P. Eng.

Poole, Jeffrey, P. Geo.
Pope, Thomas, P. Eng.
Pope, Paul F., P. Eng.
Popescu, Radu, P. Eng.
Popovic, Milorad, P. Eng.
Porter, M. Keith, P. Eng.
Porter, Neil J., P. Eng.
Pottle, Randy, P. Eng.
Powell, William G., P. Eng.
Powell, James K., P. Eng.
Powell, Kenneth W., P. Eng.
Powell, George Giles, P. Eng.
Powell, Carl, P. Eng.
Power, Adrian, P. Eng.
Power, Walter R., P. Eng.
Power, George G., P. Eng.
Power, Robert T., P. Eng.
Power, Desmond T., P. Eng.
Power, Pierce G., P. Eng.
Power, Christopher, P. Eng.
Power, Kevin C., P. Eng.
Power, D. Jason, P. Eng.
Power, Glenn R., P. Geo.
Power, Ronald, P. Eng.
Power, Douglas A., P. Eng.
Power, Sherry, P. Eng.
Power, Brian F., P. Eng.
Powers, Kenneth C., P. Eng.
Prim, Thomas M., P. Eng.
Pristach, Dusan, P. Eng.
Protulipac, Darren G., P. Eng.
Puddister, Lawrence, P. Eng.
Pugliese, Peter, P. Eng.
Pumphrey, Cyril J., P. Geo.
Pumphrey, Thomas P., P. Eng.
Quah, Ean Cheng, P. Eng.
Quaicoe, John E., P. Eng.
Quenneville, Raymond Noel, P. Eng.
Quigley, Raymond L., P. Eng.
Quinlan, Garry M., P. Geo.
Quinn, James E., P. Eng.
Quinton, Warren W., P. Eng.
Rahman, M.D. Azizur, P. Eng.
Rakowski, Wieslaw, P. Eng.
Ramadoss, Thiagarajan, P. Eng.
Rana, Madan S., P. Eng.
Randell, Walter C., P. Eng.
Randell, Charles, P. Eng.
Randell, Cordell L., P. Eng.
Randell, David B., P. Eng.
Randell, Tony, P. Eng.
Ransom, James A., P. Eng.
Raynor, Paul G. A., P. Eng.
Read, Gregory, P. Eng.
Read, Wallace S., P. Eng.
Reckling, Paul E., P. Eng.
Reed, Gordon K., P. Eng.
Rees, Alvin L., P. Eng.
Rees, D. Blair, P. Eng.
Rees, David G., P. Eng.
Reeves, David W., P. Eng.
Regular, Kevin, P. Geo.
Reid, John R., P. Eng.
Reid, J. Hugh, P. Eng.
Reid, Peter, P. Eng.
Reid, Dean, P. Eng.

Reid, Jeffrey D., P. Eng.
Reid, Paul C., P. Eng.
Reid, Elwood J., P. Eng.
Reid, Lance A., P. Eng.
Reimer, Keith Warren, P. Eng.
Rendell, Derek A., P. Eng.
Renouf, Keith, P. Eng.
Rheaume, Joel, P. Eng.
Rice, Wayne P., P. Eng.
Richards, J. David, P. Eng.
Richards, Kenneth, P. Eng.
Richards, Donald G., P. Eng.
Richards, Howard D., Jr, P. Eng.
Richardson, John, P. Eng.
Richter, Susan H., P. Eng.
Ricketts, Martin, P. Geo.
Ricks, Wayne E., P. Eng.
Rideout, John, P. Eng.
Rideout, David W., P. Eng.
Riggs, Charles J., P. Eng.
Riselli, Ronald G., P. Eng.
Ristic, Svetislav, P. Eng.
Rivera, Carlos, P. Eng.
Rixmann, D. Bradley, P. Eng.
Rizkalla, Emad, P. Eng.
Roach, William, P. Eng.
Robbins, R. Troy, P. Eng.
Robbins, Peter, P. Eng.
Robbins, Ernest A., P. Eng.
Roberts, David S., P. Eng.
Roberts, Perry H., P. Eng.
Roberts, Franklin A., P. Eng.
Roberts, Granville, P. Eng.
Roberts, Anthony, P. Eng.
Roberts, C.D. Martin, P. Eng.
Roberts, Myron Wade, P. Eng.
Robertson, H. Alexandra, P. Eng.
Robinson, Robert J., P. Eng.
Robinson, Edward G., P. Eng.
Roche, Kevin David, P. Eng.
Rochefort, Alain, P. Eng.
Rochon, Brian J., P. Eng.
Rockett, Raymond Glen, P. Eng.
Rodriguez, Rafael Mario, P. Eng.
Rodway, C.R. Wayne, P. Eng.
Rogers, Peter W., P. Eng.
Rollings, Kenneth W., P. Eng.
Rollke, Henrico, P. Eng.
Rooney, Alexander G., P. Eng.
Rose, Clifton O., P. Eng.
Rouble, Richard, P. Geo.
Rout, Richard J., P. Eng.
Rowe, Keith T.A., P. Eng.
Rowley, W. Jim, P. Eng.
Rowsell, A. Glen, P. Eng.
Rowsell, Dean F., P. Eng.
Rowsell, C. Chad, P. Eng.
Roy, Steve, P. Eng.
Royle, Patrick J., P. Eng.
Rozwadowski, Dariusz, P. Eng.
Ruelokke, Max, P. Eng.
Rundans, Valdis V., P. Eng.
Runnels, David F., P. Eng.
Russell, John H., P. Eng.
Russell, Duane Scott, P. Eng.
Rutherford, Robert J., P. Eng.

Ryan, Bernard J., P. Eng.
Ryan, Gregory, P. Eng.
Ryan, Rosalind J., P. Eng.
Ryan, Wayne, P. Eng.
Ryan, William J.J., P. Eng.
Ryan, Michael F., P. Eng.
Ryan, Joseph P., P. Eng.
Ryan, Sean, P. Eng.
Ryan, Paul, P. Eng.
Ryan, John C., P. Eng.
Ryan, Aidan F., P. Eng.
Ryder, Harry W., P. Eng.
Rypien, Donald, P. Eng.
Sabin, Dr. Gary, P. Eng.
Sacuta, Paul A., P. Eng.
Salsman, Richard D., P. Eng.
Sambotelecan, Constantin, P. Eng.
Samms, Glenn J., P. Eng.
Sanger, Andrew D., P. Eng.
Sani, Roberto, P. Eng.
Santo, Bruce, P. Eng.
Saoudy, Saoudy Ahmed, P. Eng.
Saunders, Randy, P. Eng.
Saunders, Phillip, P. Geo.
Saunders, Gregory D., P. Eng.
Saunders, Jeffrey G., P. Eng.
Saunders, Cynthia, P. Geo.
Sauvageau, Andre, P. Eng.
Savage, Gary W., P. Eng.
Sawler, Derek E., P. Eng.
Sceviour, Paul, P. Eng.
Schell, Joseph P., P. Eng.
Schillereff, H. Scott, P. Geo.
Schlereth, David P., P. Eng.
Schofield, Lloyd Arthur, P. Eng.
Schofield, Lawrence, P. Eng.
Schor, Michael A., P. Eng.
Schwartz, John T., P. Eng.
Scott, Brian E., P. Eng.
Scott, Susan A., P. Geo.
Scott, Jerry, P. Eng.
Scott, William J., P. Eng./P. Geo.
Scott, William A., P. Eng.
Scott, Robert G., P. Eng.
Seaby, Cordell S., P. Eng.
Searle, Shawn S., P. Eng.
Sears, W. Brian, P. Geo.
Sellars, Sadie L., P. Eng.
Sentell, Peter J., P. Eng.
Serrano, Luis Ernesto, P. Eng.
Seshadri, Rangaswamy, P. Eng.
Seto, John, P. Eng.
Severs, Richard, P. Eng.
Sexton, Jeffrey B., P. Eng.
Seymour, Robert N., P. Eng.
Shapter, Peter E., P. Eng.
Sharan, Anand M., P. Eng.
Sharp, R. Glenn, P. Eng.
Sharp, James J., P. Eng.
Shaw, Duncan G., P. Eng.
Sheaves, Gordon C., P. Eng.
Sheppard, Dean R., P. Geo.
Sheppard, Bernard, P. Geo.
Sheppard, David, P. Eng.
Sheppard, Eric, P. Eng.
Sheppard, Robert E., P. Eng.

Sheppard, Charles E., P. Eng.
Sheppard, Bradley T., P. Eng.
Sheps, Sidney, P. Eng.
Shirokoff, John, P. Eng.
Short, David L., P. Eng.
Short, Lloyd G.M., P. Eng.
Shortall, John M., P. Eng.
Shortall, Christopher, P. Eng.
Sich, Martin, P. Eng.
Siew, Ting M., P. Eng.
Sim, Bosco Tinyin, P. Eng.
Simmons, John L., P. Eng.
Simms, Herbert T., P. Eng.
Sinclair, Iain, P. Geo.
Singh, Vijay, P. Geo.
Singleton, L. John, P. Eng.
Singleton, John A., P. Eng.
Sinyard, G.R. Jason, P. Eng.
Sivakolundu, Gururajan, P. Eng.
Skanes, Eric R., P. Eng.
Skarborn, Stig R., P. Eng.
Skinner, Byron, P. Eng.
Skinner, Kevin, P. Eng.
Skoda, Jiri, P. Eng.
Skoll, Simon, P. Eng.
Slade, James R.M., P. Eng./P. Geo.
Slade, George Guy, P. Eng.
Sloka, Linda Jane, P. Eng.
Small, Andrew, P. Eng.
Smink, Willem K., P. Eng.
Smirnoff, Anna, P. Geo.
Smith, Harvey A., P. Geo.
Smith, Janice L., P. Geo.
Smith, Frank D., P. Eng.
Smith, Walter L., P. Eng.
Smith, G. Keith, P. Eng.
Smith, Gerald W., P. Eng.
Smith, Alexander, P. Eng.
Smith, David A., P. Eng.
Smith, Stuart D., P. Eng.
Smith, Harry A., P. Eng.
Smith, Wallace R., P. Eng.
Smith, C. Todd, P. Eng.
Smith, Kenneth W., P. Eng.
Smith, Peter J.R., P. Eng./P. Geo.
Smith, Michael John, P. Eng.
Smith, William G., P. Eng.
Smith, Peter N., P. Eng.
Smyth, Francis P., P. Eng.
Snelgrove, Lynette G., P. Eng.
Snelgrove, Terence, P. Geo.
Snook, Jeffrey, P. Eng.
Snook, Edwin J., P. Eng.
Snow, Gary, P. Geo.
Snow, William F., P. Eng.
Snyder, G. Gregory, P. Eng.
Snyder, Harold L., P. Eng.
Sodha, Narendra, P. Eng.
Somerton, Cordswell N., P. Eng.
Sooley, Trevor G., P. Eng.
Soper, Christopher, P. Eng.
Southward, Ralph, P. Eng.
Sparkes, Kerry, P. Geo.
Sparkes, John N., P. Eng.
Sparkes, Stephen, P. Eng.
Sparrow, Michael J., P. Eng.

Spencer, Michael B., P. Eng.
Spencer, Donald S., P. Eng.
Spencer, Paul Robert, P. Eng.
Spencer, David J., P. Eng.
Spencer, Garry L., P. Eng.
Spicer, David J., P. Eng.
Spracklin-Reid, Darlene L., P. Eng.
Spracklin, Derrick I., P. Eng.
Spurrell, G. Richard, P. Eng.
Squires, Robert A., P. Eng.
Squires, Calvin J., P. Eng.
Squires, Gerald C., P. Geo.
Squires, Douglas J., P. Eng.
Srinivasa, Muthu, P. Eng.
St-Arnaud, Jacques, P. Eng.
St. George, D. Seumas, P. Eng.
St. George, Kevin F., P. Eng.
Stacey, Craig, P. Eng.
Stacey, Brian D., P. Eng.
Stanford, A. Reginald, P. Eng.
Stapleton, Gerald F., P. Eng.
Stapleton, Gregory, P. Geo.
Steel, Abigail M., P. Eng.
Steele, Donald F., P. Eng.
Steels, Gordon Tomas, P. Eng.
Steeves, Kevin Carl, P. Eng.
Steeves, Allen L., P. Eng.
Stein, Cecil James, P. Eng.
Stevenson, Garry Wayne, P. Eng./P. Geo.
Stevenson, Bruce C., P. Eng.
Stewart, John McG., P. Geo.
Stockley, Glenn E., P. Eng.
Stojanovic, Miodrag, P. Eng.
Stokes, Clark, P. Eng.
Stokes, Robert E., P. Geo.
Stone, Terence, P. Eng.
Stone, Barry M., P. Eng.
Stone, Gerald S., P. Eng.
Stoodley, Raymond G., P. Eng.
Stratton, Scott J., P. Eng.
Stratton, Christopher, P. Eng.
Stratton, Carl, P. Eng.
Strickland, Tracy Lynn, P. Eng.
Strickland, Donald B., P. Eng.
Strickland, Roland E., P. Geo.
Strong, Warren E., P. Eng.
Strong, Derek C., P. Eng.
Strong, Stewart C., P. Eng.
Stuckless, Gregory, P. Eng.
Stuckless, Hugh P., P. Eng.
Sturge, L. George, P. Eng.
Sturton, Anthony R., P. Eng.
Styran, Terry P., P. Eng.
Subramanyam, Balakrishnan, P. Eng.
Suek, Gerry, P. Eng.
Sugnanam, S. Naidu, P. Eng.
Sullivan, Michael A., P. Eng.
Sultan, Babar, P. Eng.
Summers, Robin B., P. Eng.
Swamidas, Arisi, P. Eng.
Swanson, Eric, P. Geo.
Swantee, John P., P. Eng.
Swinden, H. Scott, P. Geo.
Szoo, Charles F., P. Eng.
Taite, Brian, P. Eng.
Taite, Craig J., P. Eng.

Talabany, Khasraw F., P. Eng.
Tallman, Peter, P. Geo.
Tan, Tom M.H., P. Eng.
Tancock, Robert J., P. Eng.
Tapp, Roger J., P. Eng.
Tarrant, Donald R., P. Eng.
Tay, David C., P. Eng.
Taylor, Brian B., P. Eng.
Taylor, Kenneth R., P. Eng.
Taylor, N.R., P. Eng.
Tee, Dion R., P. Eng.
Temple, S. Dale, P. Eng.
Templeton, David S., P. Eng.
Terpstra, Jelle, P. Eng.
Terreault, Pierre-Henri, P. Eng.
Thanassoulis, Pericles, P. Eng.
Thangasamy, Christian G., P. Eng.
Thibert, Francois, P. Geo.
Thielmann, Victor, P. Eng.
Thistle, Sue Ann, P. Eng.
Thistle, David, P. Eng.
Thomas, Peter W., P. Eng.
Thomas, Trevor, P. Eng.
Thomas, Karen E., P. Eng.
Thompson, David A., P. Eng.
Thompson, Brian E., P. Eng.
Thompson, Lorne W., P. Eng.
Thoms, Joseph, P. Eng.
Thomson, Barry J., P. Eng.
Thomson, Ian C. E., P. Eng.
Thorburn, Raymond S., P. Eng.
Thorburn, James P., P. Eng.
Thorburn, Robert J., P. Eng.
Thornhill, Clyde G., P. Eng.
Thurlow, J. Geoffrey, P. Geo.
Tietz, Klaus M., P. Eng.
Tiller, Richard W., P. Eng.
Tiller, Harry L., P. Eng.
Tilley, David C., P. Eng.
Tilley, Derek E., P. Eng.
Tinto, Iain A., P. Eng.
Tobiasz, Slawomir, P. Eng.
Tobin, Kenneth R., P. Eng.
Tobin, Brian T., P. Eng.
Tojcic, John, P. Eng.
Toll, Michael O., P. Eng.
Tonn, Karl F., P. Eng.
Totten, Gary, P. Eng.
Trahey, Joseph I., P. Eng.
Tran, Long T., P. Eng.
Tran, Trong C., P. Eng.
Tremblay, Martine, P. Eng.
Trenholm, Robert G., P. Eng.
Trenholm, Barry, P. Eng.
Trudeau, Marc, P. Eng.
Tu, Richard, P. Eng.
Tuach, John, P. Geo.
Tucker, Randy C., P. Eng.
Tucker, Kyle, P. Eng.
Tucker, John Richard, P. Eng.
Tucker, William L., P. Eng.
Tucker, Gary D., P. Eng.
Tucker, Bill, P. Eng.
Tulk, Donald, P. Eng.
Tupper, Allison D., P. Eng.
Turpin, Arden E., P. Eng.

Turpin, Joseph, P. Eng.
Tworzyanski, Piotr Jakub, P. Eng.
Ullah, Wasi, P. Eng.
Upshall, Peter, P. Eng.
Upshall, Darrell K., P. Eng.
Urquhart, Alexander G., P. Eng.
Valliant, Wayne, P. Geo.
Vallis, George R., P. Eng.
Valsamis, Alexandre M., P. Eng.
Van Brunt, Graham, P. Eng.
Vangool, William, P. Eng.
Vardy, William E., P. Eng.
Vardy, Darrell J., P. Eng.
Varma, Sudhanshu K., P. Eng.
Vatcher, Spencer V., P. Geo.
Vatcher, Thomas R., P. Eng.
Veilleux, Byron Wade, P. Geo.
Veitch, Brian J., P. Eng.
Venkatesan, Ramachandran, P. Eng.
Verge, Robert W., P. Eng.
Versavel, Patrick A.G., P. Eng.
Versloot, George, P. Eng.
Versteeg, Andrew W., P. Eng.
Vertolli, Augusto, P. Eng.
Vetter, William J., P. Eng.
Vickers, Edward F., P. Eng.
Vidal, Javier, P. Eng.
Vincent, Jeffrey W., P. Eng.
Vincent, Thomas G., P. Eng.
Vincent, Richard B., P. Eng.
Vincent, Kevin R., P. Eng.
Vineham, Gregory C., P. Eng.
Vivian, Cecil R., P. Eng.
Vivian, Frederick G., P. Eng.
Vokey, Gary Maxwell, P. Eng.
Vrana, Jan, P. Eng.
Wadden, David J., P. Eng.
Wade, Brian M., P. Eng.
Wadhawan, Suraj P., P. Eng.
Wadhwa, Ramendra S., P. Eng.
Wagner, Jan Vaclav, P. Eng.
Wahba, Youssef L., P. Eng.
Wakefield, Robert, P. Eng.
Wakeham, Graham, P. Eng.
Walker, Alan James, P. Eng.
Walker, Daniel L.N., P. Eng.
Walsh, Denis F., P. Geo.
Walsh, Michael J., P. Eng.
Walsh, Adrian C., P. Eng.
Walsh, Derrick J., P. Eng.
Walsh, Robert J., P. Eng.
Walsh, Brian J., P. Eng.
Walsh, Brian J., P. Eng.
Walsh, A. Gerard, P. Eng.
Walters, Barry J., P. Eng.
Wardle, Richard, P. Geo.
Wares, Roy, P. Geo.
Warford, Victoria H., P. Eng.
Warford, James C., P. Eng.
Warren, Garry B., P. Eng.
Warren, Andrew W., P. Eng.
Warren, Ivan W., P. Eng.
Warren, A. Craig, P. Eng.
Waterman, Wade Q., P. Eng.
Watson, Raymond Neil, P. Geo.
Watt, Terence W., P. Eng.

Wawrzkow, Michel T., P. Eng./P. Geo.
Way, Brent C., P. Eng.
Webster, Cory, P. Eng.
Wegenast, William G., P. Eng.
Wellon, O. Keith, P. Eng.
Wells, Stuart, P. Geo.
Wells, Geoffrey, P. Eng.
Wells, James R., P. Eng.
Wells, Darrell F., P. Eng.
Wensman, Gary L., P. Eng.
Westermann, Gerold, P. Eng.
Whalen, Philip Roy, P. Eng.
Whalen, Brian P., P. Eng.
Whalen, Anthony C., P. Eng.
Whalen, Elizabeth, P. Eng.
Whalen, Darlene, P. Eng.
Wheeler, Robert, P. Geo.
Wheeler, Randy A., P. Eng.
Whelan, J. Gerald, P. Geo.
Whelan, Madeline, P. Eng.
Whelan, Thomas G., P. Eng.
Whelan, Gary F., P. Eng.
Whelan, John R., P. Eng.
Wheller, Steven G., P. Eng.
White, Richard, P. Eng.
White, Leo H., P. Eng.
White, Gerard, P. Eng.
White, R. Keith, P. Eng.
White, George F., P. Eng.
White, Daniel G., P. Eng.
White, William L.G., P. Eng./P. Geo.
White, Arthur L., P. Eng.
White, Gervase, P. Eng.

Whitehorne, Robert G., P. Eng.
Whiteway, R. Keith, P. Eng.
Whitten, John J., P. Eng.
Whitten, Reginald E., P. Eng.
Whittle, G. Paul, P. Eng.
Whorrall, David, P. Eng.
Wilcox, Frederick, P. Eng.
Wilcox, Lawrence B., P. Eng.
Wilkinson, Graham C., P. Eng.
Will, Raymond S., P. Eng.
Willar, Glenn M.J., P. Eng.
Williams, Albert, P. Eng.
Williams, Curtis, P. Eng.
Williams, Shawn R., P. Eng.
Williams, Mark, P. Eng.
Williams, W. Gary, P. Eng.
Williamson, George F., P. Eng.
Willoughby, David, P. Eng.
Wilson, Mark R., P. Geo.
Wilson, Robert W., P. Eng.
Wilson, J. Donald, P. Eng.
Wilson, Norman, P. Eng.
Wilson, James Ian, P. Eng.
Wilson, Harrison, P. Eng.
Wilson, Mark, P. Eng.
Wilton, Derek H.C., P. Geo.
Windsor, H. Neil, P. Eng.
Windsor, Kenneth C., P. Eng.
Winsor, William D., P. Eng.
Winsor, Roderick S., P. Eng.
Winsor, Fraser N., P. Eng.
Winsor, Glenn R., P. Eng.
Wiseman, Rupert, P. Eng.

Wiseman, Robert J., P. Eng.
Wishahy, Momen A., P. Eng.
Wong, Daniel K., P. Eng.
Woodfine, Alphonsus, P. Eng.
Woodford, Edward J., P. Eng.
Woodford, Paul G., P. Eng.
Woodland, F.C. Bruce, P. Eng.
Woodman, K. Kirk, P. Geo.
Woods, Gary, P. Geo.
Woods, Dennis, P. Eng.
Woodworth-Lynas, Christopher, P. Geo.
Woody, Donald, P. Eng.
Woolgar, Susann, P. Eng.
Woolgar, Robert D., P. Eng.
Woolhouse, Thomas Henry, P. Eng.
Wright, James A., P. Geo.
Wright, Patrick J., P. Eng.
Wright, David A., P. Eng.
Yenumula, N. Prasad, P. Eng.
Yeo, Gam, P. Eng.
Yetman, Paul, P. Eng.
Yetman, Deana C., P. Eng.
Yetman, Richard D., P. Eng.
Young, Gary H., P. Eng.
Young, Craig Harvey, P. Eng.
Young, Keith G., P. Eng.
Young, Eric M., P. Eng.
Young, Harvey F., P. Eng.
Young, Peter V., P. Eng.
Zaichkowsky, Michael, P. Eng.
Zalzala, Adnan J., P. Eng.
Zevenhuizen, John, P. Geo.
Zinck, Philip A., P. Eng.

Bibliography of Works Consulted

Much more information on particular areas treated in this volume may be found by consulting the following references.

BOOKS
Alcock, Sir John; Brown, Sir Arthur Whitten, *Our Transatlantic Flight*, Kimberley and Company Limited, London, 1969
Baker, Melvin, *The Power of Commitment*, Silver Lights Club, c/o Newfoundland and Labrador Hydro, St. John's, Newfoundland, 2000
Baker, Melvin; Pitt, Robert D.; Pitt, Janet Miller, *The Illustrated History of Newfoundland Light and Power*, Creative Publishers, St. John's, Newfoundland, 1990
Barclay, James C., *Wiring the Island*, unpublished manuscript, 1995
Cardoulis, John, *A Friendly Invasion: The American Military in Newfoundland 1940-1990*, Breakwater, St. John's, Newfoundland, 1990
Clayton, Howard, *Atlantic Bridgehead*, Garnstone Press, London, England, 1968
Cranford, Garry, *The Buchans Miners*, Flanker Press Limited, St. John's, Newfoundland, 1997
Galgay, Frank; McCarthy, Michael; Okeefe, Jack, *The Voice of Generations*, RP Books, St. John's, Newfoundland, 1994
Geren, Richard; McCullogh, Blake, *Cain's Legacy*, Iron Ore Company of Canada, Sept Iles, Quebec, 1990
Hammond, Rev. John W., *The Beautiful Isles*, Pentecostal Assemblies of Newfoundland, St. John's, Newfoundland, 1978
Hierlihy, Oscar, G., *Memoirs of a Newfoundland Pioneer in Radio and Television*, Breakwater Books, St. John's, 1995
Hiller, J. K., *The Newfoundland Railway 1881-1949*, Newfoundland Historical Society Pamphlet Number 6, St. John's, 1981

House, Edgar, *Edward Feild - The Man and His Legacy*, Jesperson Printing Limited, St. John's, Newfoundland, 1987

Janzen, Olaf, *Putting the Hum on the Humber*, Corner Brook Pulp and Paper Limited, Corner Brook, Newfoundland, 1999

Lingard, Mont, *The Newfie Bullet*, Mont Lingard Publishing, Grand Falls-Windsor, 2000

Maynard, Lara, editor, *Hibernia: Promise of Rock and Sea*, Breakwater Books, St. John's, Newfoundland, 1998

Martin, Wendy, *Once upon a Mine: Story of Pre-confederation Mines on the Island of Newfoundland*, Special Volume 26, The Canadian Institute of Mining and Metallurgy, Montreal, Quebec, 1983

MacLeod, Malcolm, *A Bridge Built Halfway - A History of Memorial University College, 1925-1950*, McGill-Queens University Press, Montreal, 1990

MacLeod, Malcolm, *Nearer than Neighbours*, Harry Cuff Publications Limited, St. John's, Newfoundland, 1982

McAllister, R. I., *Newfoundland and Labrador, the First Fifteen Years of Confederation*, Dicks and Co. Ltd., St. John's, Newfoundland, 1966

Mott, Henry Youmans, *Newfoundland Men*, T.W. and J. F. Cragg, Concord, New Hampshire, 1894

Neary, Steve, *The Enemy on our Doorstep*, Jesperson Press, St. John's, Newfoundland, 1994

Newfoundland Historic Trust, *A Gift of Heritage: Historic Architecture of St. John's*, Valhalla Press, Canada, 1975

Newfoundland Statistics Agency, Government of Newfoundland and Labrador, *Historical Statistics of Newfoundland and Labrador Volume II (VII)*, St. John's, Newfoundland, November 1994

Ozorak, Paul, *Abandoned Military Installations in Canada, Volume 3:Atlantic*, 2001

Penney, A. R., *A History of the Newfoundland Railway Volume 1 (1881-1923)*, Harry Cuff Publications Ltd., St. John's, Newfoundland, 1988

Penney, A. R., *A History of the Newfoundland Railway Volume II (1923-1988)*, Harry Cuff Publications Limited, St. John's, Newfoundland, 1990

Prowse, D. W., *History of Newfoundland*, Macmillan and Co., London and New York, 1895

Rowe, Frederick W., *The History of Education in Newfoundland*, Ryerson Press, Toronto, 1952

Rowe, Frederick W., *Education and Culture in Newfoundland*, McGraw-Hill Ryerson Limited, Scarborough, Ontario, 1976

Rowe, Frederick W., *A History of Newfoundland and Labrador*, McGraw Hill-Ryerson Limited, Toronto, 1980

Smallwood, Joseph R., *Books of Newfoundland*, Newfoundland Book Publishers Limited, St. John's, Newfoundland, 1975, 1979

Smallwood, Joseph R., *Encyclopedia of Newfoundland and Labrador*, Newfoundland Book Publishers (1967) Limited, St. John's, Newfoundland, 1981

Smith, Philip, Brinco: *The Story of Churchill Falls*, McClelland and Stewart Limited, Toronto, Ontario, 1975

Sparkes, Paul, *A History of Canadian National Telecommunications in Newfoundland*, unpublished manuscript, 1976

Steele, Donald H., editor, *Early Science in Newfoundland and Labrador*, Avalon Chapter of Sigma Xi, St. John's, Newfoundland, 1987

Swain, Hector K., *VOWR: The Unfolding Dream*, Creative Publishers, St. John's, Newfoundland, 1999

Swanson, E. A.; Strong, D. F.; Thurlow, J. G., editors, *The Buchans Orebodies: Fifty Years of Geology and Mining*, Geological Association of Canada, Waterloo, Ontario, 1981

Tarrant, D. R., *Atlantic Sentinel*, Flanker Press, St. John's, Newfoundland, 1999

Tarrant, D. R., *Marconi's Miracle*, Flanker Press, St. John's, Newfoundland, 2001

Weir, Gail, *The Miners of Wabana*, Breakwater Books, St. John's, Newfoundland, 1989

PERIODICALS, PUBLICATIONS and PAPERS

Annual Reports, from various companies

Atlantic Oil and Gas Week, Halifax, Nova Scotia, August 2001

CBC Radio Commemorative Brochure 1932-1992, John Power,
    Judy Squires, John O'Mara, St. John's, 1993

Dialogue for Engineers and Geoscientists, Association of Professional
    Engineers and Geoscientists of Newfoundland,
    St. John's, various issues

*Daily News (*newspaper*)*, St. John's, Newfoundland, various issues

The Economy, publication of the Economics and Statistics Branch
    (Economic Research and Analyses), Department of Finance,
    Government of Newfoundland and Labrador, 2000, 2001,
    St. John's, Newfoundland

Flagship, publication of Terra Nova Development, St. John's,
    Newfoundland, July/August 2001

Grand Falls News Special Anniversary Edition, Abitibi-Price,
    Grand Falls, 1984

Highway to Progress, publication of the province of Newfoundland, 1966

Newfoundland Radio in Pictures, A. B. Sullivan et al, 1952

Oil and Gas Report, Department of Mines and Energy, Government
    of Newfoundland and Labrador, June 2001

*The (Evening) Telegram (*newspaper*)*, St. John's, Newfoundland, various
    issues

Twenty-fifth Anniversary, Association of Professional Engineers
    of Newfoundland, St. John's, 1977

Index

Abery, C. 77
Abitibi-Consolidated 160, 165, 166
Acres Canadian Bechtel (ACB) 76
Admiralty House 33
AETTN 233
Agra-BAE Newplan Joint Venture 79
Aguathuna 56, 148
airports 116
Alcock, Sir J. 112-113
Alexis bridge 87,90
alphabet fleet 109
Aluminum Company of Canada 141
American Newfoundland Fluorspar 140
Angel, F. W. 133
Anglo-American Telegraph Company 14-19, 37, 39
Anglo-Newfoundland Development Corporation 58, 62, 103, 134-136, 138, 157-161
APEGN (APEN) 3-4, 231-258
APEGN award recipients 243-247
APEGN charter members 233
APEGN former presidents 236-239
APEGN list of members 248-258
APEGN registrations (1997-2001) 235
aquaculture 194-195
Argentia 20, 43, 47, 56, 105, 110, 116, 150, 173, 175-179, 186, 188, 242
Argentia Naval Base 20, 43, 105, 116, 175-179, 186, 188
Armstrong Whitworth 161
ASARCO 135-138
asbestos 148
Ash, E. 39

AT&T 27-28, 31-32
Atlantic Coast Copper Corporation 127
Atlantic Telegraph Company 13
Avalon Cablevision 47
Avalon Telephone 18, 23-31, 41, 44, 46
aviation 111-117

Baie Verte 42, 60, 127, 148-149
Bailey, A. 61
Baker, H. A. 121
Barrelman 43
Bay d'Espoir 5, 60, 65-70, 77-78
Bay of Islands Power Company 58-59
Bay Roberts 17-18, 26, 40, 56, 90
Beeton, M. 157
Bell Canada 20, 23, 26, 30
Bell Island 4, 5, 24, 26, 40, 56, 61-62, 110, 120, 127-133, 148, 173, 229, 241
Ben Nevis 211,219
Bennett, C. F. 123
Betts Cove 99, 125-127
Birdseye, C. 192
Bishop's Falls 18, 25, 58, 93, 103, 105-106, 160
Bishop's Falls mill 160-61
Black River 156
Blackman A. L. 95, 99-101
Bond, Sir R. 102, 157
Botwood 18, 20, 25, 47, 58, 103, 107,113-115, 137, 157-160, 173
Botwood Railway 103
Botwood seaplane base 113-114
Bowater 18. 25, 58-59, 62, 65-66,163

Bowater Power Company 59
Bowater, Sir E. 163
Brait, A. A. 26, 27, 30
bridges 90-94
British Newfoundland Development Corporation 65-66, 71-72, 75-76, 225
broadcasting 37-48
Brown, Sir A. 112-113
Bruneau, A. 63, 224
Buchans 4-5, 18, 22, 58, 77, 103, 105, 134-138, 232
Buchans Mining Company 18, 22, 136-137
Buchans Railway 136
buildings 169-173
Bull Arm 6, 9, 14, 206-207, 209- 210, 215, 217
Bullet (train) 106
Burchell, H. C., 173
Burry, Rev. L. 42

Cable and Wireless 16, 32
Cable Atlantic 47
cable TV 47
Cameron, A. 63
Camp Alexander 179-181
Campbellton mill 160
Canadian Broadcasting Corporation 42, 44-47
Canadian International Paper Company 163
Canadian Marconi 25, 32, 38-39
Canadian National Railways 20, 94, 105-107, 109-110, 202
Canadian National Telecommunications 20-23, 27
Canadian Overseas Telecommunications Corporation 31-32
Canning, S. 12
Canning, W. 134
Cape Race 13, 24, 30, 34
Cape Race Loran C 34

Cape Ray 11-12, 34, 37
Carbonear 11, 12, 24, 42, 55, 103, 195, 229
Carew, Dr. S. J. 129
Caribou 110
Carol Lake 143,144
Carter, Sir F. 97
Castle Hill 2
Cat Arm hydro development 78
Catalina 22, 57
Cataracts bridge 91-92
causeways 89-90
CBPP 163-164
Centre for Cold Ocean Resources Engineering 225-226
Centre for Earth Resources Research 226
Chalker, J. 77
Chambers, R. E. 128
Churchill Falls Labrador Corporation 70-76
Churchill Falls waterfall 50
Clarenville 27-28, 30, 32, 42, 57, 86, 88, 103, 196, 229
Clarenville Light and Power 57
CN Marine 110
Cochrane, Governor 84
cod fishery 193-194
College of the North Atlantic 229
Collins, J. 38
Come by Chance 5, 67, 98, 202-204
Commercial Cable Company 16-17, 19
Commission of Government 86, 121, 183-184, 196
communications 8-48, 186-188
communications and artillery sites 186-188
Conception Bay Electric Company 55
Concrete Products 88
construction 169-188
copper 99, 119-120, 122-127, 134, 137-138, 186, 201

Corner Brook 4, 23, 25, 30, 43-44, 56, 58-60, 67-69, 86, 105-106, 162-163, 173
Corner Brook paper mill 161-165
Cow Head 198-199
Crosbie, J. 77

Deer Lake 32, 45, 47, 56, 58-60, 67, 77, 86, 88, 116, 137, 162-165
Deer Lake power station 59
Delaney, J. 18

Department of Posts and Telegraphs 19-20
Desbarats, G. 64-65
Direct United States Cable Company 16
dolomite 145, 150
Dominion Bridge 94
Dominion Broadcasting Company 41-42
Dominion Iron and Steel Company 129-130
Dominion Steel and Coal 18, 130-131, 133, 140, 148
Donich, J. 23
Doyle, J. 146, 166

Earhart, A. 113
early electrical utilities 53-60
early engineering 1-7
early mining 122-124
early wireless 34
Eastern Telephone and Telegraph 27,32
Edmund B. Alexander 43, 168, 179
Electric Reduction Company of Canada 66, 71
electricity 51-81
electricity, other providers 57-60
Ellershausen, Baron F. von 125-126
Engineering Institute of Canada 3, 227

ExxonMobil Canada 213

FENCO 93,174
Fessenden, R. 37
Field, C. 11-14
fish freezing 192
fish salting 192
Fisheries and Marine Institute 226-227
Fisher's mill 167
fluorspar 138, 139, 140,141
Fort McAndrew 177
Fort Pepperrell 179-181
Fort Townshend 2,84
Fortis 31, 63
FPSO 213-218
Frequency conversion 68-69
Friede Goldman 198-199, 218
Fullerton, D 77
future oil developments 218-219
Future SET 234-235

Gander 20, 42, 44, 46-47, 62, 88, 98, 114-116, 175-176, 183-185, 188, 229
Gander base 183-185
Gander International Airport 115
Geological Survey of Newfoundland 120-122
Gisborne, F. W. 9-12, 123-124
Glovertown mill 161
gold 120, 125, 138, 150-151
Golden Eagle 202
Goose Bay 20, 26-27, 43, 45-46, 65, 76, 78, 89, 91, 110, 116, 148, 166, 175-176, 184-186, 188
Goose Bay base 185-186
Gosse, J. 23
Grand Banks oil 205-217
Grand Falls 4, 18, 25-27, 42, 44-47, 57-58, 67-70, 86,103, 107, 134, 156-161, 188, 229, 241
Grand Falls paper mill 156-160

Grand Lake 69
Granite Canal 78
Great Eastern 9, 14-15
Groom, D. 76-77

Harbour Grace 15-17, 24, 52, 55, 99, 101, 103, 113, 191, 195-196
Harbour Grace Gas Light Company 52
Harmon Air Force Base 157,176, 181- 183, 186, 188
Harmsworth H. 158
Harmsworth, A. 158
Harvey and Company 156, 192
Heart's Content 9, 14-17, 39, 55-57, 103, 158, 195
Hibernia 5, 197, 201, 206-213, 218, 234, 241
Hierlihy, O. 40-41, 44, 46
highways 83-90
Hobbs, G. 65, 67
Hollinger Company 142
Holyrood 5, 67, 70-71, 101, 202
Holyrood refinery 202
Hope Brook 150
Howley, J. 121
Howse, C. K. 121
Husky Oil 213, 218

Imperial and International Communications Company 16
INCO 242
Installed generating capacity 81
Institute for Marine Dynamics 227-228
International Power and Paper Company 137, 162-163
iron 4-5, 60, 70, 72, 107, 119-120, 127-133, 141-143, 145-148, 151, 210
Iron Ore Company of Canada 28, 47, 58, 60, 72, 142-146, 150

Jamieson, D. 44-45
Joyce Rev. J. 38-39

Knob Lake 142-143
Kruger 163-164

La Manche 122-123
Labrador City 5, 28, 46, 60, 89, 108, 141-146
Labrador Mining and Exploration Company 142, 146
Laima 115
Lavino, E. 140
lead 4, 66, 120, 122-123, 134-135, 186
limestone 148-150, 152
Lindbergh C. 112
Little Bay 99, 122, 126-127
local telephone service 18-31
Low, A. P. 142
Lucky Strike mine 135-136
Lundberg, H. T. 135-136

MacDonald, D. 77
Mackay, A. M. 37, 53-54
Mackay, W. 49
Manuels 149-150
Marconi, G. 10, 34-37
marine 189-199
Maritime Mining Corporation 125
Marystown 42, 47, 67, 94, 196,-199, 218-219
Marystown Shipyard 195-199, 218-219
Massey, G. 49
McLean, J. 70
McNamara Construction 77-78, 88, 93
McParland, Donald 73-75
Memorial University of Newfoundland 6, 45, 129, 171-172, 221-227, 234-235
Menihek 37, 53

Michel, M. 134-135
microwave systems 32
military bases 175-188
mineral shipments 152
mining 119-152
Mobil
Mobile Big Pond 61
Monkman, B. 232-233, 236
Montreal Engineering 55, 60, 68
Morris, S. H. 26
Murphy R.J. 23-26
Murphy, J. J. 23-24, 55
Murray, A. 97, 120-121

NC-4 111-112
Newdock 199
New York, Newfoundland and London Telegraph Company 12, 15
Newfoundland and Labrador Hydro chairs 77
Newfoundland and Labrador Hydro-electric Corporation 32, 60, 63, 70, 76-78
Newfoundland and Labrador Power Corporation 76
Newfoundland Base Contractors 180, 182
Newfoundland Broadcasting Company 41, 44, 46
Newfoundland Consolidated Copper Mining Company 126-127
Newfoundland Electric Telegraph Company 11-12
Newfoundland Fluorspar Limited 18, 141
Newfoundland Government Railway 104
Newfoundland (Light and) Power 32, 55, 57, 59-64, 223
Newfoundland Postal Telegraphs 19
Newfoundland Power and Paper Company 161

Newfoundland Power Commission 60, 64-76
Newfoundland Power Commission chairs 65
Newfoundland Products Corporation 58-60, 161
Newfoundland Railway 4, 20, 55, 94-95, 103-105, 109, 136
Newfoundland Telephone 23-31
Newfoundland Transshipment 197, 217-218
North Atlantic Refining 203-204
North Limited 146
North West River cable car 91
Northcliffe, Lord 156-158
Nova Scotia Telegraph Company 11
Nugget Pond 151

Ocean Engineering Research Centre 226
Ocean Ranger 206, 210
oil and gas 201-219
Osberg, G. 26

Paradise River hydro development 78, 79
Parsons Pond 205-205
Pepperrell 43, 168, 175, 179-182, 186, 188
Peter Kiewit Sons 197-198, 219
Petro-Canada 204, 211, 213
Petty Harbour hydro plant 4, 49, 54-55, 61
phosphorous 71, 132-133
Placentia lift bridge 93
Port aux Basques 5, 19-20, 25, 30, 37, 45, 47, 56, 67, 88, 102-103, 105-108, 110, 173, 229
port development 173-175
Port Union 57, 190, 196
power grid 5, 60, 67, 78
Price (Newfoundland) Pulp and Paper Limited 159-160

Price Company limited 138
provincial grid 5, 60, 67, 77
Public Service Electric Company 56, 63
pulp and paper 155-167
pulp, Newfoundland's first mill 156
pyrophyllite 149

Quebec North Shore and Labrador Railway 56, 63, 108, 143-144
Queen of Swansea 124

radio 37-46
railroad spur routes 103
railroads 3, 83, 94-108, 136
Rattling Brook 62
Read, A. C. 112
Read, Wallace 76-77
recent electrical utilities 60-81
Reed Company 58, 159-160
refineries 67, 202-205
Regional Cablesystems 47
Reid Newfoundland Company 19, 39, 54-55, 102-103, 107-109, 134, 161
Reid, Sir R. 49, 54, 98, 101-102, 109
Retty, Dr. J. 142
Rio Tinto 65, 73, 146
Rogers Cantel 30
Rothermere mine 137-138
Rothermere, Lord 156, 158-159
Royal Canadian Air Force 20, 43, 86, 115, 180, 184-186
Royal Securities 60
Ryan, A. 63

satellite TV 48
Schefferville 60, 72, 107, 142-143
Scotia Pier 128, 130-132
Scott, W. 134
Scully mine 146
Seibert, W. E. 139-140

Shaheen, J. 202-203
Shawinigan Engineering 88, 144
Shawmont Newfoundland Engineering 68
Shawmont Resources 70
Shellbird Cablevision Company 47
shipbuilding 195-199
Shoal Bay 120
Shoe Cove satellite station 28
significant producing mines and quarries 150
silicosis 140
Smallwood, J. R. 43, 65-66, 73, 86, 88, 110, 145, 166, 202-203
Snelgrove, A. K. 121
splinter fleet 109
Springdale 60, 67, 69
St. George's 47, 56-57, 88, 91, 97-99, 150, 181, 196, 205
St. John's 2, 3, 5, 10-13, 15-20, 22-28, 30-31, 34, 37-38, 40, 42-47, 49, 51-56, 60-62, 67, 70, 77, 82, 84, 86-88, 90, 93-94, 96-99, 101-108, 112, 116-117, 120, 122, 124,133, 139, 141, 149, 156-158, 160, 168-169, 170- 176, 179-181, 192, 194-195, 206, 213, 219-221, 226, 228-229, 231-232
St. John's and Carbonear Electric Telegraph 11
St. John's city engineers 172
St. John's dry dock 104-105
St. John's Electric Light Company 53- 54, 60-61
St. John's Gas Light Company 51-52
St. John's harbour development 174-175
St. John's Light and Power 55
St. John's Railway Station 95
St. John's Street Car Company 49, 104
St. John's Street Railway 54-55
St. Lawrence 5, 18, 56, 138-141

St. Lawrence Corporation of Newfoundland 139-141
Stairs, D. 62-63
Stephenville 20, 23, 44, 47, 56-57, 66-67, 77, 116, 165-166, 166, 175, 181-183, 186
Stephenville paper mill 165-166
Stirling, G. 44
Stott, D. 19
streetcars 54, 82
Stuewe, W. 131, 232-233
Symonds, R. 23

telegraph communications 4, 6, 7, 9-18
telephone communications 18-32
Telesat Canada 28
television 46-47
Templeton, D. 62-63
Terra Nova FPSO 213, 215-217
Terra Nova oil 213-217
Terra Nova Telecommunications 20-23, 30
Terra Transport 106
Thorburn, R. 99, 101
Tilt Cove 99, 124-125, 127
Tilt Cove Copper Company
Tors Cove power station 61, 64
transatlantic telegraph cables 13-18
transatlantic telephone cables 31-32
transatlantic voice communications 31-32
Trans-Canada Highway 5, 30, 86-88, 93, 106
trans-island railway 97
trans-island telegraph line 11-13
Trans-Labrador Highway 89-91
transportation 83-117
troposcatter systems 28, 186-188
turret 213, 215, 216
Twillingate Telephone and Electric Company 18, 22
Twin Falls 58, 60, 72, 144

Union Electric Light and Power Company 57, 63
Union Mining Company 124
United Towns Electric Company 18, 25, 55-57, 63, 66
Unitel Newfoundland 31
US Military 20, 43, 173, 175-188

Verran, H. 123
vocational schools 228-229
Voisey's Bay 234, 242

Wabana Light and Power 56
Wabush 5, 28, 45, 58, 60, 72-73, 76-77, 116, 144, 146-148, 150, 166
Wabush Mines 58, 60, 144, 146-148, 150, 167
wartime airport construction 116
west coast oil 204-205
West Coast Power 56, 63
Western Union Telegraph Company 17-18, 22-23
Whalen, N. 77
Whalesback 69, 127
Whiffen Head 211, 217-218
Whitbourne 19, 99, 101, 188
Whiteway, Sir W. 96, 99, 101-103
William Carson 110
Williams, Dr. H. 224
wireless communications 34-37
World War Two attacks 130

Young, V. 77

Zanussi 117
zinc 4, 134, -135, 139